FOOD INSECURITY
and Hunger in the United States

AN ASSESSMENT OF THE MEASURE

Panel to Review the U.S. Department of Agriculture's Measurement of
Food Insecurity and Hunger

Gooloo S. Wunderlich and Janet L. Norwood, *Editors*

Committee on National Statistics

Division of Behavioral and Social Sciences and Education

NATIONAL RESEARCH COUNCIL
OF THE NATIONAL ACADEMIES

THE NATIONAL ACADEMIES PRESS
Washington, D.C.
www.nap.edu

THE NATIONAL ACADEMIES PRESS 500 Fifth Street, N.W. Washington, DC 20001

NOTICE: The project that is the subject of this report was approved by the Governing Board of the National Research Council, whose members are drawn from the councils of the National Academy of Sciences, the National Academy of Engineering, and the Institute of Medicine. The members of the committee responsible for the report were chosen for their special competences and with regard for appropriate balance.

This study was supported by Contract/Grant No. 43-3AEM-3-80125 between the National Academy of Sciences and the U.S. Department of Agriculture. Support of the work of the Committee on National Statistics is provided by a consortium of federal agencies through a grant from the National Science Foundation (Number SBR-0112521). Any opinions, findings, conclusions, or recommendations expressed in this publication are those of the author(s) and do not necessarily reflect the views of the organizations or agencies that provided support for the project.

Library of Congress Cataloging-in-Publication Data

Food insecurity and hunger in the United States : an assessment of the measure / Panel to Review U.S. Department of Agriculture's Measurement of Food Insecurity and Hunger ; Gooloo S. Wunderlich and Janet L. Norwood, editors.
    p. cm.
  Includes bibliographical references and index.
    ISBN 0-309-10132-8 (pbk.) — ISBN 0-309-65805-5 (pdfs) 1. Food supply— United States. 2. Hunger—United States. I. Wunderlich, Gooloo S. II. Norwood, Janet Lippe. III. National Research Council (U.S.) Panel to Review U.S. Department of Agriculture's Measurement of Food Insecurity and Hunger.
  TX360.U6F677 2006
  363.80973—dc22
                                     2006005691

Additional copies of this report are available from National Academies Press, 500 Fifth Street, N.W., Lockbox 285, Washington, DC 20055; (800) 624-6242 or (202) 334-3313 (in the Washington metropolitan area); Internet, http://www.nap.edu

Printed in the United States of America

Suggested citation: National Research Council. (2006). *Food Insecurity and Hunger in the United States: An Assessment of the Measure.* Panel to Review the U.S. Department of Agriculture's Measurement of Food Insecurity and Hunger, Gooloo S. Wunderlich and Janet L. Norwood, *Editors,* Committee on National Statistics, Division of Behavioral and Social Sciences and Education. Washington, DC: The National Academies Press.

THE NATIONAL ACADEMIES
Advisers to the Nation on Science, Engineering, and Medicine

The **National Academy of Sciences** is a private, nonprofit, self-perpetuating society of distinguished scholars engaged in scientific and engineering research, dedicated to the furtherance of science and technology and to their use for the general welfare. Upon the authority of the charter granted to it by the Congress in 1863, the Academy has a mandate that requires it to advise the federal government on scientific and technical matters. Dr. Ralph J. Cicerone is president of the National Academy of Sciences.

The **National Academy of Engineering** was established in 1964, under the charter of the National Academy of Sciences, as a parallel organization of outstanding engineers. It is autonomous in its administration and in the selection of its members, sharing with the National Academy of Sciences the responsibility for advising the federal government. The National Academy of Engineering also sponsors engineering programs aimed at meeting national needs, encourages education and research, and recognizes the superior achievements of engineers. Dr. Wm. A. Wulf is president of the National Academy of Engineering.

The **Institute of Medicine** was established in 1970 by the National Academy of Sciences to secure the services of eminent members of appropriate professions in the examination of policy matters pertaining to the health of the public. The Institute acts under the responsibility given to the National Academy of Sciences by its congressional charter to be an adviser to the federal government and, upon its own initiative, to identify issues of medical care, research, and education. Dr. Harvey V. Fineberg is president of the Institute of Medicine.

The **National Research Council** was organized by the National Academy of Sciences in 1916 to associate the broad community of science and technology with the Academy's purposes of furthering knowledge and advising the federal government. Functioning in accordance with general policies determined by the Academy, the Council has become the principal operating agency of both the National Academy of Sciences and the National Academy of Engineering in providing services to the government, the public, and the scientific and engineering communities. The Council is administered jointly by both Academies and the Institute of Medicine. Dr. Ralph J. Cicerone and Dr. Wm. A. Wulf are chair and vice chair, respectively, of the National Research Council.

www.national-academies.org

Acknowledgments

T he Panel to Review the U.S. Department of Agriculture's Measurement of Food Insecurity and Hunger acknowledges with appreciation the contributions of the many persons who gave generously of their time and knowledge to this study.

Support for the study was provided by the Economic Research Service (ERS) and the Food and Nutrition Service (FNS) of the U.S. Department of Agriculture (USDA). We particularly wish to thank Mark Prell, chief, Food Assistance Branch, who served as project officer of the study. He and his colleagues, particularly Mark Nord and Margaret Andrews, were helpful in providing information about the research undertaken in the development and implementation of the measures of food insecurity and hunger and answering the many questions from the panel and staff. We acknowledge Susan Offutt, administrator, and Phil Fulton, former associate administrator of ERS; Steven Carlson, director, family programs staff, Office of Analysis, Nutrition, and Evaluation, Food and Nutrition Service; Betsey Kuhn, director of the Food Economics Division (FED), ERS; and David Smallwood, deputy director for food assistance and nutrition research, FED, for recognizing the need and initiating and supporting this important study.

In addition, we acknowledge the many federal and nonfederal government officials and those from the research and academic community who gave expert presentations to the panel at its initial meeting and participated in the workshop on the measurement of food insecurity and hunger. We are grateful to the authors of the papers prepared for the workshop. Their names and the subjects of their papers are listed in Chapter 1. The papers were used by the panel and staff to guide them in drafting this report.

A number of people in the Committee on National Statistics (CNSTAT) and the Division of Behavioral and Social Sciences and Education (DBASSE) provided support and assistance to the study panel. We acknowledge with gratitude the contributions of the staff. The panel wishes to thank Shelly Ver Ploeg, who as study director until October 1, 2004, was chiefly responsible for developing and organizing the workshop during the first phase of the study. The panel also appreciates the fine work of Gooloo Wunderlich who was responsible for preparing the drafts of the reports and responding to the many comments from the reviewers on behalf of the panel. Lance Hunter handled administrative matters. Throughout, the panel benefited from the advice and collaboration provided by Connie Citro, CNSTAT director. Christine McShane, senior editor of the DBASSE reports office, provided professional editing advice, and Kirsten Sampson Snyder efficiently shepherded the report through the report review and production process. In addition we would like to thank Linda Meyers, director, Board on Food and Nutrition, Institute of Medicine, who made available the 1990 legislation and early documents in the development of the food security measure and for being available throughout to answer questions related to the development of the measure and nutrition.

Finally, I would like to thank the members of the panel for their generous contributions of time and expert knowledge to the deliberations and preparation of this report.

This report has been reviewed in draft form by individuals chosen for their diverse perspectives and technical expertise, in accordance with procedures approved by the National Research Council's Report Review Committee. The purpose of this independent review is to provide candid and critical comments that will assist the institution in making its published report as sound as possible and to ensure that the report meets institutional standards for objectivity, evidence, and responsiveness to the study charge. The review comments and draft manuscript remain confidential to protect the integrity of the deliberative process. We thank the following individuals for their review of this report: Jay Bhattacharya, Stanford Medical School, Stanford University; Peter Eisinger, Department of Urban Politics and Economic Development Policy, Wayne State University; Jean-Pierre Habicht, Division of Nutritional Sciences, Cornell University; William L. Hamilton, Abt Associates, Cambridge, MA; William D. Kalsbeek, Department of Biostatistics and Survey Research Unit, University of North Carolina; Valerie Tarasuk, Department of Nutritional Sciences, University of Toronto; Howard Wainer, Measurement Consulting, National Board of Medical Examiners, Philadelphia; and Catherine E. Woteki, Scientific Affairs, Mars, Incorporated, McLean, VA.

Although the reviewers listed above have provided many constructive comments and suggestions, they were not asked to endorse the conclusions or recommendations nor did they see the final draft of the report before its release. The review of this report was overseen by David M. Betson, Department of Economics and Policy Studies, Notre Dame University, and John C. Bailar III, Department of Health Studies (emeritus), University of Chicago. Appointed by the National Research Council, they were responsible for making certain that an independent examination of this report was carried out in accordance with institutional procedures and that all review comments were carefully considered. Responsibility for the final content of this report rests entirely with the authoring committee and the institution.

Janet L. Norwood, *Chair*
Panel to Review the U.S. Department of
Agriculture's Measurement of Food
Insecurity and Hunger

Contents

Executive Summary

The United States is viewed by the world as a country with plenty of food, yet not all households in America are food secure, meaning access at all times to enough food for an active, healthy life. A proportion of the population experiences food insecurity at some time in a given year because of food deprivation and lack of access to food due to economic resource constraints. Still, food insecurity in the United States is not of the same intensity as in some developing countries.

Since 1995 the U.S. Department of Agriculture (USDA) has annually published statistics on the extent of food insecurity and food insecurity with hunger in U.S. households. These estimates are based on a survey measure developed by the U.S. Food Security Measurement Project, an ongoing collaboration among federal agencies, academic researchers, and private organizations. It is an experiential measure based on reported behaviors, experiences, and conditions in response to questions in a household survey. The measure was developed over the course of several years in response to the National Nutrition Monitoring and Related Research Act of 1990 (NNMRR). The legislation specifically called for development of a standardized mechanism and instrument(s) for defining and obtaining data on the prevalence of food insecurity in the United States and methodologies that can be used across the NNMRR programs and at the state and local levels.

The USDA estimates of food insecurity are based on data collected annually in the Food Security Supplement (FSS) to the Current Population Survey (CPS). On the basis of the number of food-insecure conditions reported, households are classified into one of three categories for purposes of monitoring and statistical analysis of the food security of the U.S. population: (1) *food secure,* (2) *food insecure without hunger,* and (3) *food insecure with hunger.*

The USDA estimates, published in a series of annual reports, are widely used by government agencies, the media, and advocacy groups to report the extent of food insecurity and hunger in the United States, to monitor progress toward national objectives, to evaluate the impact of particular public policies and programs, as a standard by which the performance of USDA programs is measured, and as a basis for a diverse body of research relating to food assistance programs.

In addition, USDA has a program of research for improving the measurement and understanding of food security. Despite these efforts, some major questions continue to be raised regarding the underlying concepts, the estimation methods, and the design and clarity of the questions used to construct the food insecurity scale.

PANEL CHARGE

USDA requested the Committee on National Statistics of the National Academies to convene a panel of experts to undertake a two-year study in two phases to review at this 10-year mark the concepts and methodology for measuring food insecurity and hunger and the uses of the measures. The specific tasks to be addressed in Phase 1 of the study were:

- the appropriateness of a household survey as a vehicle for monitoring on a regular basis the prevalence of food insecurity among the general population and within broad population subgroups, including measuring frequency and duration;
- the appropriateness of identifying hunger as a severe range of food insecurity in such a survey-based measurement method;
- the appropriateness, in principle and in application, of item response theory and the Rasch model as a statistical basis for measuring food insecurity;
- the appropriateness of the threshold scores that demarcate food insecurity categories—particularly the categories "food insecure with hunger" and "food insecure with hunger among children"—and the labeling and interpretation of each category;
- the applicability of the current measure of the prevalence of food insecurity with hunger for assessing the effectiveness of USDA food assistance programs, in connection with the Government Performance and Results Act performance goals for the Food and Nutrition Service; and

- future directions to consider for strengthening measures of hunger prevalence for monitoring, evaluation, and related research purposes.

In Phase 2 of the study the panel was to consider in more depth the issues raised in Phase 1 relating to the concepts and methods used to measure food security and make recommendations as appropriate. In addition, the panel was asked to address and make recommendations on:

- the content of the 18 items and the set of food security scales based on them currently used by USDA to measure food insecurity;
- how best to incorporate and represent information about food security of both adults and children at the household level;
- how best to incorporate information on food insecurity in prevalence measures;
- needs and priorities for developing separate, tailored food security scales for population subgroups, for example, households versus individuals, all individuals versus children, and the general population versus homeless persons; and
- future directions to consider for strengthening measures of food insecurity prevalence for monitoring, evaluation, and related research purposes throughout the national nutrition monitoring system.

The Committee on National Statistics appointed a panel of 10 experts to examine the above issues. In order to provide timely guidance to USDA, the panel issued an interim Phase 1 report, *Measuring Food Insecurity and Hunger: Phase 1 Report*. That report presented the panel's preliminary assessments of the food security concepts and definitions; the appropriateness of identifying hunger as a severe range of food insecurity in such a survey-based measurement method; questions for measuring these concepts; and the appropriateness of a household survey for regularly monitoring food security in the U.S. population. It provided interim guidance for the continued production of the food security estimates. This final report primarily focuses on the Phase 2 charge. The major findings and conclusions based on the panel's review and deliberations are summarized below, followed by the text of all of the recommendations.

FINDINGS AND CONCLUSIONS

Concepts and Definitions

The broad conceptual definitions of food security and insecurity developed by the expert panel convened in 1989 by the Life Sciences Research Office (LSRO) of the Federation of American Societies for Experimental

Biology have served as the basis for the standardized operational definitions used for estimating food security in the United States. *Food security* according to the LSRO definition means access at all times to enough food for an active, healthy life. *Food insecurity* exists whenever the availability of nutritionally adequate and safe foods or the ability to acquire acceptable foods in socially acceptable ways is limited or uncertain. Food insecurity as measured in the United States refers to the social and economic problem of lack of food due to resource or other constraints, not voluntary fasting or dieting or because of illness or for other reasons. Although lack of economic resources is the most common constraint, food insecurity can also be experienced when food is available and accessible but cannot be used because of physical or other constraints, such as limited physical functioning by elderly people or those with disabilities.

Food insecurity is measured as a household-level concept that refers to uncertain, insufficient, or unacceptable availability, access, or utilization of food. It is therefore households that are classified as food secure or food insecure. It means that one can measure and report the number of people who are in food-insecure households (even though not everyone in the household need be food insecure themselves). When a household contains one or more food-insecure persons, the household is considered food insecure.

A full understanding of food insecurity requires the incorporation of its frequency and duration because more frequent or longer duration of periods of food insecurity indicate a more serious problem. *Frequency and duration are therefore important elements for USDA to consider in the concept, operational definition, and measurement of household food insecurity and individual hunger.*

The LSRO conceptual definition of hunger adopted by the interagency group on food security measurement is: "The uneasy or painful sensation caused by a lack of food, the recurrent and involuntary lack of access to food. Hunger may produce malnutrition over time. . . . Hunger . . . is a potential, although not necessary, consequence of food insecurity" (Anderson, 1990, pp. 1575, 1576). This language does not provide a clear conceptual basis for what hunger should mean as part of the measurement of food insecurity. The first phrase "the uneasy or painful sensation caused by a lack of food" refers to a possible consequence of food insecurity. The second phrase "the recurrent and involuntary lack of access to food" refers to the whole problem of food insecurity, the social and economic problem of lack of food as defined above.

Unlike food insecurity, which is a household-level concept, hunger is an individual-level concept. The Household Food Security Survey Module (HFSSM) in the Food Security Supplement to the CPS measures food insecurity at the household level; it does not measure the condition of hunger at the individual level. The HFSSM does include items that are related to being hungry. Some or all of these items are probably appropriate in the food insecurity scale, but they contribute to the measurement of food insecurity and not the measurement of hunger.

The panel therefore concludes that hunger is a concept distinct from food insecurity, which is an indicator of and possible consequence of food insecurity, that can be useful in characterizing severity of food insecurity. Hunger itself is an important concept that should be measured at the individual level distinct from, but in the context of, food insecurity.

The broad conceptual definition of household food insecurity includes more elements than are included in the current USDA measure of food insecurity. Not all elements of the consensus conceptual definition of food insecurity have been incorporated into the USDA measurement of food insecurity in the United States. It was a decision of the Food Security Measurement Project to limit the operational definition and measurement approach to only those aspects of food insecurity that can be captured in a household-level survey. The other conceptually separable aspects of food insecurity are potentially distinct empirical dimensions. For example, the measurement does not include the supply of food or its safety or nutritional quality; these additional aspects would require developing measures and fielding separate surveys to measure them. Moreover, the food supply in the United States is generally regarded as safe, and nutritional adequacy is already assessed by other elements of the nutrition monitoring system, in particular the continuing National Health and Nutrition Examination Survey. *The panel therefore concludes that it is neither required nor necessarily appropriate for USDA to attempt to measure all the elements of the broad conceptual definition of food insecurity as part of the HFSSM.*

The labeling used to categorize food insecurity is at the heart of the criticism of the current measurement system. In particular, the category "food insecure with hunger" has come under scrutiny because of disagreement over whether hunger is actually measured. The rationale for including hunger in the label for the classification is understandable. Hunger is a politically sensitive and evocative concept that conjures images of severe deprivation, and the HFSSM does include some items that are specifically related to hunger. However, the measurement of food insecurity rather than hunger is the primary focus of the HFSSM. As an indication of the severity of food insecurity, the HFSSM asks the household respondent if in the past 12 months she or he has experienced being hungry because of lack of food due to resource constraints. This is not the same as evaluating individual members of the household in a survey as to whether or not they have experienced hunger. *The panel urges USDA to consider alternate labels to convey the severity of food insecurity without the problems inherent in the current labels.*

Survey Measurement of Food Insecurity and Hunger

The panel reviewed the current questions used to measure food insecurity and hunger, considered the relationship among the three major aspects of food insecurity and hunger embodied in the questions (whether the household experienced uncertainty, the perception of insufficiency in quality of diet, and reduced food intake or the feeling of hunger), and identified several design issues in the HFSSM that should be addressed.

USDA's food security scale measures the severity of food insecurity in surveyed households and classifies their food security status during the previous year. The frequency of food insecurity and the duration of spells of insecurity are not assessed directly in the HFSSM questions that are used to classify households by food security status. Although some of the response options do offer the choice of "often, sometimes, or never," these response options are not sufficient measures of frequency, and they are not included in the construction of the scale. In addition to the items in the HFSSM, the full supplement includes questions that focus on duration. However, these questions are not part of the 18-item HFSSM, although they have been used in research to estimate the percentage of the population that is food insecure on a given day in a given month. A recent study undertaken by USDA researchers examined the extent to which food insecurity and hunger are occasional, recurring, or frequent in the U.S. households that experience them. *The panel recommends further research on the frequency and duration of food insecurity.*

The panel reviewed the 18 items that constitute the food insecurity scales as well as the entire questionnaire module in which these are embedded and found many issues of questionnaire design. Consistent terminology, clustering questions so as to focus on a specific reference person or reference group (e.g., the respondent, all adults in the household, all children in the household) and on a specific reference period (e.g., 12 months versus 30 days), and developing response options that most closely map to the respondent's representation of the behavior or attitude are all means by which questions can be designed to reduce cognitive burden and thereby improve the validity and reliability of the measures. Inevitably, questionnaire design requires balancing multiple intents and principles, and there is no perfect questionnaire design. Nevertheless, the panel concludes that the questions in the HFSSM in particular and the FSS in general can be improved by attending to these design principles to the extent possible.

Item Response Theory

USDA uses the Rasch model, a specific type of item response theory (IRT) model, to estimate the food insecurity of households. Several issues have been raised about the use of IRT models in the measurement of food insecurity and, in particular, the use of the Rasch model.

The panel reviewed IRT and related statistical models and discussed their use and applicability to the development of such classifications as food insecurity. The panel recommends modifications of the current IRT methodology used by USDA to increase the amount of information that is used and to make the methodology more appropriate to the types of data that are currently collected using the Food Security Supplement to the Current Population Survey.

The panel reviewed how the latent variable models are estimated and issues of identifiability of these models and how IRT models are used by USDA in the measurement of food insecurity. On the basis of this review, the panel suggests how the models might be used in better ways to accomplish this measurement and recommends a simple way to modify the existing models currently used by USDA to take into account the polytomous nature of the data collected.

Survey Vehicles to Measure Food Insecurity and Hunger

USDA bases its annual report and estimates of the prevalence of food insecurity on data collected from the Food Security Supplement to the Current Population Survey. The Household Food Security Survey Module, or a modification of it, is or has been used in several surveys. One of the main objectives of the annual food insecurity measure is to monitor the estimated prevalence of food insecurity, as well as changes in its prevalence over time, at the national and state levels to assess both program policies and the possible need for program development.

After reviewing the key features of selected national surveys—the Current Population Survey, the National Health Interview Survey, the National Health and Nutrition Examination Survey, and the Survey of Income and Program Participation—the panel compared the relative merits of each, for either carrying the Food Security Supplement or conducting research to supplement the information obtained from it. The panel recommends research and testing to understand better the strengths and weaknesses of each survey in relation to the Current Population Survey, leading to the selection of a specific survey vehicle for the Food Security Supplement, or for supplementing that information for research purposes.

Food Insecurity Estimates as a Measure of Program Performance Assessment

Currently, the Food and Nutrition Service in USDA uses trends in the prevalence of food insecurity with hunger based on the HFSSM as a measure of its annual performance to implement the Government Performance and Results Act of 1993 (GPRA). That law requires government agencies to account for progress toward intended results of their activities. It requires that specific performance goals be established and that annual measurement of these output goals be undertaken to determine the success or failure of the program. The panel was asked to comment on the applicability of these data for this purpose.

The panel concludes that an overall national estimate of food insecurity is not appropriate as a measure for meeting the requirements of the GPRA. Even an appropriate measure of food insecurity or hunger using appropriate samples would not be a useful performance indicator of food assistance programs, because their performance is only one of many factors that result in food insecurity or hunger. Consequently, changes in food insecurity and hunger could be due to many factors other than the performance of the food safety net.

The panel concludes that relying exclusively on trends in prevalence estimates of food insecurity as an indicator of program results is inappropriate. To assess program results, a better understanding is needed of the transitions into and out of poverty made by low-income households and the kind of unexpected changes that frequently bring about alterations—for good or bad—in households participating in food assistance programs.

Conclusion

The panel is impressed with the extensive research thus far undertaken, and with the continuing research carried out by USDA. The panel urges that the research program be continued and makes several recommendations for its direction in the future.

The panel concludes that the measurement both of food insecurity and of hunger is important. The recommendations in the report are intended to improve these measurements, so that policy makers and the public can be better informed. Toward this end, the panel has recommended research efforts that should lead to improved concepts, definitions, and measurement of food insecurity and hunger. The panel has provided a detailed discussion of the analytical methods used by USDA and made recommendations for further research to improve the accuracy of the food insecurity scale and on survey alternatives. The panel recognizes that such research will take time.

RECOMMENDATIONS

On the basis of its findings and conclusions, the panel presents recommendations in five areas: concepts and definitions, labeling of food insecurity data outcomes, survey measurement, item response theory and food insecurity, and survey vehicles to measure food insecurity and hunger. The text of the recommendations, grouped according to these areas, follows, keyed to the chapter in which they appear in the body of the report.

Concepts and Definitions

Recommendation 3-1: USDA should continue to measure and monitor food insecurity regularly in a household survey. Given that hunger is a separate concept from food insecurity, USDA should undertake a program to measure hunger, which is an important potential consequence of food insecurity.

Recommendation 3-2: To measure hunger, which is an individual and not a household construct, USDA should develop measures for individuals on the basis of a structured research program, and develop and implement a modified or new data gathering mechanism. The first step should be to develop an operationally feasible concept and definition of hunger.

Recommendation 3-3: USDA should examine in its research program ways to measure other potential, closely linked, consequences of food insecurity, in addition to hunger, such as feelings of deprivation and alienation, distress, and adverse family and social interaction.

Recommendation 3-4: USDA should examine alternate labels to convey the severity of food insecurity without the problems inherent in the current labels. Furthermore, USDA should explicitly state in its annual reports that the data presented in the report are estimates of prevalence of household food insecurity and not prevalence of hunger among individuals.

Survey Measurement

Recommendation 4-1: USDA should determine the best way to measure frequency and duration of household food insecurity. Any revised or additional measures should be appropriately tested before implementing them in the Household Food Security Survey Module.

Recommendation 4-2: USDA should revise the wording and ordering of the questions in the Household Food Security Survey Module. Examples of possible revisions that should be considered include improvements in the consistent treatment of reference periods, reference units, and response options across questions. The revised questions should reflect modern cognitive questionnaire design principles and new data collection technology and should be tested prior to implementation.

continued on next page

Item Response Theory and Food Insecurity

Recommendation 5-1: USDA should consider more flexible alternatives to the dichotomous Rasch model, the latent variable model that underlies the current food insecurity classification scheme. The alternatives should reflect the types of data collected in the Food Security Supplement. Alternative models that should be formally compared include:

- Modeling ordered polytomous item responses by ordered polytomous rather than dichotomized item response functions.
- Treating items with frequency follow-up questions appropriately, for example, as a single ordered polytomous item rather than as two independent questions.
- Allowing the item discrimination parameters to differ from item to item when indicated by relevant data.

Recommendation 5-2: USDA should undertake the following additional analyses in the development of the underlying latent variable model:

- Fitting models that allow for different latent distributions for households with children and those without children and possibly other subgroups of respondents.
- Fitting models that allow for different item parameters for households with and without children for the questions that are appropriate for all households in order to study the possibility and effects of differential item functioning.
- Studying the stability of the measurement system over time, possibly using the methods of differential item functioning.

Recommendation 5-3: To implement the underlying latent variable model that results from the recommended research, USDA should develop a new classification system that reflects the measurement error inherent in latent variable models. This can be accomplished by classifying households probabilistically along the latent scale, as opposed to the current practice of deterministically using the observed number of affirmations. Furthermore, the new classification system should be more closely tied to the content and location of food insecurity items along the latent scale.

Recommendation 5-4: USDA should study the differences between the current classification system and the new system, possibly leading to a simple approximation to the new classification system for use in surveys and field studies.

Recommendation 5-5: USDA should consider collecting data on the duration of spells of food insecurity in addition to the currently measured intensity and frequency measures. Measures of frequency and duration spells may be used independently of the latent variable measuring food insecurity.

Survey Vehicles to Measure Food Insecurity and Hunger

Recommendation 6-1: USDA should continue to collaborate with the National Center for Health Statistics to use the National Health and Nutrition Examination Survey to conduct research on methods of measuring household food insecurity and individual hunger and the consequences for nutritional intake and other relevant health measures.

Recommendation 6-2: USDA should carefully review the strengths and weakness of the National Health Interview Survey in relation to the Current Population Survey in order to determine the best possible survey vehicle for the Food Security Supplement at a future date. In the meantime, the Food Security Supplement should continue to be conducted in the Current Population Survey.

Recommendation 6-3: USDA should explore the feasibility of funding a one-time panel study, preferably using the Survey of Income and Program Participation, to establish the relationship between household food insecurity and individual hunger and how they co-evolve with income and health.

1

Introduction

The United States is considered the land of abundant food and most Americans are food secure, meaning access at all times to enough food for an active, healthy life. A proportion of the population experiences food insecurity at some time in a given year, however, because of food deprivation and lack of access to food due to economic resource constraints. Still, food insecurity in the United States is not of the same intensity as in some developing countries. The U.S. Department of Agriculture (USDA) estimated that in 2004 11.9 percent of U.S. households were food insecure at some time during the year. That means they did not have access at all times to enough food or were uncertain of having or were unable to acquire enough food for all household members because of insufficient economic or other resources. Of these, 3.9 percent of the households were estimated by USDA as "food insecure with hunger," that is, food insecurity in the household reached levels of severity great enough that one or more household members were hungry at least some time during the year because they could not afford enough food (Nord, Andrews, and Carlson, 2005b, pp. 4–5). Such prevalence of food insecurity has economic and public health consequences for both the individuals and their communities as a result of reduced cognitive development and learning capacity in children, as well as lower intakes of food energy and key food nutrients and other similar conditions.

The statistics on food insecurity and hunger in U.S. households, published annually by USDA, are based on a survey measure developed by the U.S. Food Security Measurement Project, an ongoing collaboration among federal agencies, academic researchers, and private organizations. The measure was developed over the course of several years in response to the National Nutrition Monitoring and Related Research Act of 1990. It is a direct experiential measure based on self-reported behaviors, experiences, and conditions in response to questions in a survey. (A brief history of the development of the measure is provided in Chapter 2.)

Each year since 1995, USDA has developed annual estimates of the prevalence of food insecurity for U.S. households. These estimates are developed using data collected annually in the Food Security Supplement (FSS) to the Current Population Survey (CPS). On the basis of the number of food-insecure conditions that are reported by households (i.e., the number of questions the respondent affirms), USDA classifies households into one of three categories for purposes of monitoring and statistical analysis of the food security of the population: food secure, food insecure without hunger, and food insecure with hunger. Furthermore, the questions specify that the behavior or condition must be due to a lack of economic or other resources to obtain food, so the scale is not affected by hunger due to voluntary dieting or fasting or being too busy to eat or other similar reasons, or involuntary hunger due to reasons other than resource constraints. USDA uses statistical methods based on a single-parameter logistic item response theory model (the Rasch model) to assess individual questions and to assess the assumptions that justify using the raw number of items affirmed as an ordinal measure of food insecurity. (This method and the issues surrounding its use for this purpose are described in detail in Chapter 5.)

The USDA estimates, published in a series of annual reports since 1995, are widely used by government agencies, the media, and advocacy groups to report the extent of food insecurity and hunger in the United States, to monitor progress toward national objectives, to evaluate the impact of particular public policies and programs, as a standard by which the performance of USDA programs is measured, and as a basis for a diverse body of research on questions related to food assistance programs. Government agencies have also adopted the estimates as targets for performance assessment. The U.S. Department of Health and Human Services (DHHS) uses the food security measure to assess the performance of its Healthy People 2010 initiative. The Food and Nutrition Service of USDA is using the measure as a target for its strategic plan to fulfill requirements of the Government Performance and Results Act of 1993.[1]

---

[1] "The Government Performance and Results Act of 1993 seeks to shift the focus of government decision making and accountability away from a preoccupation with the activities that are undertaken, such as grants dispensed or inspections made, to a focus on the results of those activities, such as real gains in employability, safety, responsiveness, or program quality. Under the act, agencies are to develop multiyear strategic plans, annual performance plans, and annual reports" (U.S. Government Accountability Office, 2002, p. 1).

Despite the extensive use of the measure over the years, some major questions continue to be raised related to the concepts, methods, and questionnaire items used by USDA for measuring food insecurity and hunger in the annual surveys.

While the USDA annual reports define the concepts of food insecurity and the three categories of food insecurity that are estimated and reported (i.e., food secure, food insecure without hunger, and food insecure with hunger), providing detail about how they are measured, the terms "food security" and "food insecurity" are relatively new to both policy makers and the public, and they are sometimes confusing. While the term "hunger" is not new, measurement of hunger and how hunger fits conceptually into food insecurity is not completely clear. As currently construed in USDA's food insecurity measure, hunger could be considered a severe level of food insecurity. This use of the term "hunger" has been questioned by some who believe that hunger is conceptually distinct from food insecurity. Because the label "hunger" is a politically potent concept, the methods used to classify households as food insecure with hunger and the use of these estimates are particularly important.

Methodological and technical issues about the measure of food insecurity generally concern the appropriateness of the statistical model used in developing the food insecurity scale and the clarity and design of the CPS survey questions. Also of concern is the relatively long reference period, mixing questions focused on households with those on individuals, and using the same module to assess food insecurity among subgroups, such as households with and without children and the elderly population.

Questions about the appropriate uses of the estimates of food insecurity have also been raised. The media, advocacy groups, and others often interpret the prevalence estimates in language inconsistent with USDA usage. The primary use of the Food Security Supplement is to estimate the prevalence of food security and its severity levels. USDA has explicitly stated (Nord, Andrews, and Carlson, 2005a, p. 10) that:

> technically the Food Security Supplement data do not support estimates of the number of people that experience hunger. USDA's food security reports, based on the FSS, do not provide, nor claim to provide, statistics on the prevalence of hunger among individuals. The survey, and USDA's reports based on it, provide upper and lower bound estimates of the number of adults and number of children who were hungry at times during the year. They also provide information that sheds light on the prevalence of hunger—by describing the experiential-behavioral context in which hunger occurs. (In early years of the Food Security Measurement Project, USDA analysts sometimes used wording such as "the prevalence of hunger" as shorthand for 'the prevalence of food insecurity with hunger' in official reports and research articles. In more recent years disciplined attention has

been given to avoid such statements because of a growing awareness of the conceptual and interpretive problems they can cause.)

This understanding is consistent with the Life Sciences Research Office conceptual definitions (which grew out of considerable public discussion of a wide range of definitional/conceptual alternatives), in the sense that hunger for an individual is a potential, although not necessary, consequence of food insecurity. Yet the media and advocacy groups often interpret the prevalence estimates in language inconsistent with USDA usage.

OBJECTIVES OF THE MEASURE

The measurement and monitoring activities related to the development of the measurement of food insecurity have a number of policy-related objectives:

- Create a measure with generally agreed-on concepts, definitions, and measurement methodologies that estimates the frequency and severity of problems regarding access to food in a way that is standard and consistent over time and across subgroups of the population at national and state levels.
- Provide objective, standardized information on the extent and severity of food insecurity and the characteristics of affected persons, so that allocation of public resources and development of public policies and programs can be based on informed public debate. The mission statement of the Food and Nutrition Service (FNS), which administers USDA's food assistance programs, includes the goal of increasing food security: "FNS increases food security and reduces hunger in partnership with cooperating organizations by providing children and low-income people access to food, a healthful diet, and nutrition education in a manner that supports American agriculture and inspires public confidence."
- Provide data on household food security that can be used along with other survey information collected in surveys to assess the need for and effectiveness of public programs, especially food assistance programs; the causes of food insecurity at various levels of severity; and the effects of food insecurity on nutrition, health, children's development, and other aspects of well-being.
- Provide measures of food security that can be used consistently across the National Nutrition Monitoring and Related Research Program and in state, local, and special population surveys that can be compared meaningfully with national food security statistics.

THE PANEL'S STUDY

As indicated above, the USDA's food security measures were designed a decade ago in partnership with DHHS. USDA decided that a thorough review at this 10-year mark is warranted, especially in light of persistent conceptual and methodological concerns about the concepts and their measurement. USDA's Economic Research Service, through its Food Assistance and Nutrition Research Program, has expressed the need for a review of the conceptualization and methods used in measuring food insecurity, as well as the validity and utility of the measure for informing public policy. Promotion of food security is part of the mission of USDA's Food and Nutrition Service, and certain food security measures constitute performance goals for that agency as required by the Government Performance and Results Act.

Panel Charge

USDA requested the Committee on National Statistics of the National Academies to convene a panel of experts to provide an independent review of the current conceptualization and methods of measuring food insecurity and hunger in the U.S. population. The contract charge to the panel specifies that the 2-year study will be conducted in two phases. During Phase 1 of the study a workshop was to be held to address the key issues laid out for the study and a short report prepared based on workshop discussions and preliminary deliberations of the panel. The specific tasks to be addressed in Phase 1 include:

- the appropriateness of a household survey as a vehicle for monitoring on a regular basis the prevalence of food insecurity among the general population and within broad population subgroups, including measuring frequency and duration;
- the appropriateness of identifying hunger as a severe range of food insecurity in such a survey-based measurement method;
- the appropriateness, in principle and in application, of item response theory and the Rasch model as a statistical basis for measuring food insecurity;
- the appropriateness of the threshold scores that demarcate food insecurity categories—particularly the categories "food insecure with hunger" and "food insecure with hunger among children"—and the labeling and interpretation of each category;
- the applicability of the current measure of the prevalence of food insecurity with hunger for assessing the effectiveness of USDA's food assistance programs, in connection with the performance goals

pursuant to the Government Performance and Results Act (Public Law 103-62) for the Food and Nutrition Service; and

- future directions to consider for strengthening measures of hunger prevalence for monitoring, evaluation, and related research purposes.

In Phase 2 of the study the panel was to consider in more depth the issues identified in Phase 1 relating to the concepts and methods used to measure food security and make recommendations as appropriate. In addition, the panel was asked to address and make recommendations on:

- the content of the 18 items and the set of food security scales based on them currently used by USDA to measure food insecurity;
- how best to incorporate and represent information about food security of both adults and children at the household level;
- how best to incorporate information on frequency and duration of food insecurity in prevalence measures;
- needs and priorities for developing separate, tailored food security scales for population subgroups, for example, households versus individuals, all individuals versus children, and the general population versus homeless persons; and
- future directions to consider for strengthening measures of food insecurity prevalence for monitoring, evaluation, and related research purposes throughout the national nutrition monitoring system.

To address this two-phase request, the Committee on National Statistics appointed a panel of 10 members representing a range of expertise related to the scope of the study.

Study Approach

During the first phase of the study, the panel reviewed articles and papers prepared or sponsored by USDA to assess the methodological concerns about the food security measures and other published and unpublished papers.

The panel met on two occasions to deliberate on the issues listed above. The first meeting was held in March 2004. In the public part of the meeting, USDA staff and other experts in the field briefed the panel on the history of the conceptual and technical development of the measure and on the uses of the food security measure. Critics of the current measurement methodology presented their views, and USDA staff and other meeting attendees were given the opportunity to respond.

The panel held a large workshop to obtain input from a wide range of researchers and other interested members of the public. The Workshop on the Measurement of Food Insecurity and Hunger was held on July 15, 2004.

Four background papers were prepared by experts and presented at the workshop (the full text of the papers is available at http//www.national academies.org/cnstat):

- *Conceptualization and Instrumentation of Food Security* by J.P. Habicht, G. Pelto, E.A. Frongillo, and D. Rose.
- *The Uses and Purposes of the USDA Food Security and Hunger Measure* by P. Wilde.
- *Item-Response Models and Their Use in Measuring Food Security and Hunger* by M.S. Johnson.
- *Alternative Construction of a Food Security and Hunger Measure from 1995 Current Population Survey Food Security Supplement Data* by K. Alaimo and A. Froelich.

Discussants were asked to give their reactions to these papers, and open discussion sessions were set aside for general comments from participants. A roundtable discussion on the questionnaire design and cognitive aspects of the survey module was also held during the workshop.

In order to provide timely guidance to USDA, the panel issued an interim Phase 1 report, *Measuring Food Insecurity and Hunger: Phase 1 Report*. That report presented the panel's preliminary assessment of the food security concepts and definitions, the appropriateness of identifying hunger as a severe range of food insecurity in such a survey-based measurement method, questions for measuring these concepts, and the appropriateness of a household survey for regularly monitoring food security in the U.S. population. It provided interim guidance for the continued production of the food security estimates. This assessment was based primarily on the review of the literature, public comments at the panel meetings, workshop presentations and discussions, and expert judgments of the panel.

For its Phase 2 research, the panel commissioned four additional background papers from experts in the areas of hunger, methods, cognitive aspects in questionnaire development, and comparison of selected surveys to provide expert and detailed analysis of some of the key issues beyond the time and resources of its members. These papers and their respective authors are listed below.

- *Methodological Issues in Measuring Food Insecurity and Hunger* by M.S. Johnson.
- *Cognitive Aspects of the Questions Used to Measure Food Insecurity and Hunger* by J. Dykema and N.C. Schaeffer.
- *A Comparison of Surveys for Food Insecurity and Hunger Measurement* by S.J. Haider.

- *The Concept and Definition of Hunger and Its Relationship to Food Insecurity* by D.H. Holben.

The full text of these papers is available at <http://www.national acdemies.org/cnstat>.

Scope and Limitations of the Study

The study is complex, covering wide-ranging issues from concepts and definitions, to survey design and implementation, to statistical methods and models, to application for assessing program performance. At the same time, its scope is limited to reviewing USDA's measure of food insecurity as used in the annual survey. Many other issues relevant to the subject of food insecurity, such as the determinants and consequences of hunger, the relationship of obesity and food insecurity and hunger, the relationship between food insecurity and dietary intake, nutrient availability and health status, socially acceptable sources of food, and food safety, are important. These issues, however, are not intrinsically indicators of economic deprivation. Although the panel recognizes their importance, their measurement is beyond the scope of this study. Moreover, a full consideration of these issues should be the subject of separate studies.

As with most national household population surveys, the CPS excludes homeless people who are not in shelters. However, the question of including or excluding homeless people from the Food Security Supplement to the CPS is not as straightforward as for other household surveys. Omitting the homeless is likely to result in an undercount of the number of more severely food-insecure persons. The panel recognizes the likelihood of relatively high rates of food insecurity among homeless people, and the resulting negative bias resulting from their exclusion. At the same time, it has serious questions about the operational and methodological issues. Over the years, techniques have been developed to locate, sample, and obtain data about segments of this population. The Census Bureau has done a lot of experimentation in this area. Yet locating and screening respondents for eligibility require special efforts involving careful and long-term planning, substantial staff resources, considerable time, and high levels of funding. Much research and testing are required to develop the necessary protocols and procedures for conducting the Food Security Supplement in a separate survey among homeless people. Until better methods to survey the homeless are developed, continuing to limit the target population to the household population seems appropriate.

ORGANIZATION OF THE REPORT

The panel used three criteria to guide the content of the report and its recommendations. First, the subject area examined must be relevant to and within the scope and purview of the panel's charge. Second, the evidence and analysis must be sufficient to support and justify its conclusions and recommendations. Third, recommendations should be attainable at reasonable cost.

The report focuses primarily on the Phase 2 charge and is organized in a manner responsive to the panel's charge. Following this Introduction, Chapter 2 summarizes the history of the development of the concepts of household food insecurity and hunger and their operational definitions, the measurement and monitoring of food insecurity in the context of the United States using the Food Security Supplement to the Current Population Survey, and the uses of the food insecurity questions in other surveys nationally and internationally. Chapter 3 discusses the conceptual issues associated with the terms food insecurity and hunger and the operational definitions used to measure household food insecurity and hunger and the usage of labels for categories of food insecurity.

Chapters 4–7 examine a range of issues and needed changes leading toward improved measures of the prevalence of food insecurity and hunger. Chapter 4 reviews the current measurement of food insecurity and the validity and reliability of the questions used to measure food insecurity and hunger, identifying selected questions in the Household Food Security Survey Module that need improvements. Chapter 5 reviews the history and structure of latent variable models and describes the different ways of estimating latent variable models. The chapter then examines the method currently used by USDA to measure food insecurity and its prevalence and the various issues involved with the method used, suggesting better ways to match the item response theory models with the nature of the data collected in the food insecurity surveys. Chapter 6 reviews the key features of selected national sample surveys in terms of their capacity to include the Food Security Supplement, compares the relative advantages and disadvantages of the surveys, and provides recommendations for USDA's future consideration. Chapter 7 examines the use of the estimate of the prevalence of food insecurity for assessing the performance of USDA's food assistance programs in accordance with the Government Performance and Results Act of 1993.

Finally, Chapter 8 highlights the panel's key conclusions. It emphasizes that the panel, in providing the critique and recommended actions for the future, recognizes the continuing research and concerted efforts to develop a standardized direct measure of food insecurity that can be used for monitoring purposes and related research that have been carried out by USDA and its collaborating agencies.

The panel hopes that the report and its recommendations will contribute to the development of a revised, efficient, and cost-effective system for monitoring the prevalence of food insecurity in the United States, as well as provide the basis for research to answer the important questions about the broader health and socioeconomic and psychological consequences of food insecurity.

2

History of the Development of Food Insecurity and Hunger Measures

Prior to the development of the current standardized measure of the prevalence of household food insecurity in 1995, estimates of the prevalence of lack of access to food varied widely and there was little consensus over which measure was most accurate. Unlike many developing countries with widespread chronic food insecurity because of general food scarcity, food insecurity and hunger in the United States occur in a land of abundant food. Food insecurity exists in a small proportion of the population, and a smaller proportion experience hunger at some time during a year because they cannot afford enough food (LeBlanc, Kuhn, and Blaylock, 2005).

This chapter summarizes the history of the development of the measure of food insecurity in the United States from the late 1960s to the present. The chapter highlights some of the main events that have shaped the dialogue and outcomes over nearly four decades. These include government initiatives—both executive and legislative—as well as efforts of private researchers and organizations.

EARLY EFFORTS TO DEFINE HUNGER

"Hunger became a truly public issue in the United States in the late 1960s, even though a number of major federal assistance programs were already in place. The crucial period during which the issue emerged was bracketed by the April 1967 visit to the Mississippi Delta by the Senate Subcommittee on Employment, Manpower and Poverty led by Joseph Clark (D-Pa.) and Robert Kennedy (D-N.Y.), and the broadcast on May 28, 1968, of the CBS television documentary, 'Hunger in America' " (Eisinger, 1998, p. 12). The recognition that hunger exists in the United States led to an increase of federal programs and projects to eliminate the effects of poverty (Eisinger, 1996). Since the late 1960s, government agencies, academic researchers, nonprofit organizations, and advocacy groups have undertaken many studies to define and measure hunger in the American context, but without any consensus on the definition of hunger or its measurement strategy or estimates of the extent of the problem. As Radimer and colleagues observed (Radimer, Olson, and Campbell, 1990, p. 1545), "The definitions of hunger varied widely and measures of hunger were generally indirect and the definitions and measures often lacked congruence."

The term "hunger" was often used interchangeably with malnutrition, and medical and dietary intake data were used to measure the problem. Other studies attempted to use poverty data or trends in the number of people seeking food assistance as proxies for hunger. Still others attempted to gather data through various surveys. This discordance at times was a product of competing professional and political agendas (Eisinger, 1996, 1998).

THE 1980s:
THE PRESIDENT'S TASK FORCE ON FOOD ASSISTANCE

This public attention to hunger led to increases in programs and projects to alleviate the condition. In the early 1980s, adverse economic conditions and efforts to limit federal spending led to a general belief that hunger was widespread in the United States and may have been increasing. This concern led President Reagan to establish a task force to examine the food assistance programs and the claims of a resurgence of hunger. The Task Force on Food Assistance concluded that the issue of hunger was complex and observed that the terms "hunger," "poverty," and "unemployment" were often used interchangeably although they are distinct problems. Also, the population that relied on food assistance was not a homogeneous group.

After much investigative work, the task force made a distinction between two different working definitions of "hunger": (1) a scientific, clinical definition in which hunger means "the actual physiological effects of extended nutritional deprivations" and (2) a definition of hunger as commonly defined, relating more to a social phenomenon than medical results, in which hunger is "the inability, even occasionally, to obtain adequate food and nourishment. In this sense of the term, hunger can be said to be present even when there are no clinical symptoms of deprivation" (U.S. President, Task Force on Food Assistance, p. 34).

The task force concluded that "with this possible exception [the homeless], there is no evidence that widespread undernutrition is a major health problem in the United States." Yet it did find evidence of hunger "as commonly defined" (p. 36):

> To many people hunger means not just symptoms that can be diagnosed by a physician, it bespeaks the existence of a social, not a medical, problem: a situation in which someone cannot obtain an adequate amount of food, even if the shortage is not prolonged enough to cause health problems. It is the experience of being unsatisfied, of not getting enough to eat. This, of course, is the sense in which people ordinarily use the word. It is also the sense in which the witnesses before us and many of the reports and documents we have studied have spoken of hunger. This alternative definition of hunger relates directly to our communal commitment to ensure that everyone has adequate access to food, and to the nation's endeavors to provide food assistance. And in this sense we cannot doubt that there is hunger in America.

The task force, further noting in its report the lack of a definition of hunger and of documentation of it in the United States, articulated the need for measuring hunger (pp. 37, 39):

> There is no official "hunger count" to estimate the number of hungry people, and so there are no hard data available to estimate the extent of hunger directly. Those who argue that hunger is widespread and growing rely on indirect measures. . . . We regret our inability to document the degree of hunger caused by income limitations, for such lack of definitive, quantitative proof contributes to a climate in which policy discussions become unhelpfully heated and unsubstantiated assertions are then substituted for hard information.

After the 1984 task force report, researchers in the private sector and government agencies increased their efforts to develop survey measures of the severity and extent of hunger in the United States. The Food Research and Action Center sponsored a major series of surveys—the Community Childhood Hunger Identification Project (CCHIP)—to study hunger among children. The research of Radimer, Olson, Campbell, and colleagues at the Cornell University Division of Nutritional Sciences was directed toward developing indicators to assess hunger (Radimer, Olson, Green, Campbell, and Habicht, 1992; Radimer et al., 1990). At the federal government level in the mid 1980s, the U.S Department of Agriculture (USDA) began to analyze information obtained from the single survey question on the adequacy of household food supplies added since 1977 to its periodic Nationwide Food Consumption Survey. In the late 1980s, a food sufficiency question similar to the one in the Nationwide Food Consumption Surveys, along with other questions on regular access to food supplies adapted from the CCHIP questionnaire, were included by the National Center for Health Statistics (NCHS) in the third National Health and Nutrition Examination Survey (NHANES III).

THE 1990s: A PERIOD OF TRANSITION

The year 1990 marks the beginning of the emergence of consensus on the appropriate conceptual basis for defining and measuring hunger in the United States. In 1990, the Life Sciences Research Office (LSRO) of the Federation of American Societies for Experimental Biology prepared a report based on discussions of the ad hoc Expert Panel convened in 1989 on Core Indicators of Nutritional State for Difficult-to-Sample Populations for the American Institute of Nutrition, under the provisions of a cooperative agreement with the U.S. Department of Health and Human Services (DHHS). This report was published in the *Journal of Nutrition* (Anderson, 1990). It summarizes the discussions of an ad hoc expert panel charged with identifying core indicators to assess the nutritional status of difficult-to-sample populations. The report contains what have become the consensus conceptual definitions for the terms "food security," "food insecurity," and "hunger," as relevant to the United States and notes the relationship of food insecurity to hunger and malnutrition (Anderson, 1990, pp. 1575–1576, 1598).

- *Food security* was defined by the expert panel as "access by all people at all times to enough food for an active, healthy life, and includes, at a minimum: (a) the ready availability of nutritionally adequate and safe foods and (b) an assured ability to acquire acceptable foods in socially acceptable ways (e.g., without resorting to emergency food supplies, scavenging, stealing, or other coping strategies)."
- *Food insecurity* exists whenever there is "limited or uncertain availability of nutritionally adequate and safe foods or limited or uncertain ability to acquire acceptable foods in socially acceptable ways."
- *Hunger* in its meaning of "the uneasy or painful sensation caused by a lack of food" is in this definition "a potential, although not necessary, consequence of food insecurity. Malnutrition is also a potential though not necessary consequence of food insecurity. . . . Hunger, as a recurrent and involuntary lack of access to food which may produce malnutrition over time, is discussed as food insecurity in this report."

Lack of a standard operational definition of hunger in the past had been a major obstacle in estimating the extent of the problem. The expert panel decided that redefining the hunger problem in terms of food security could overcome some of these problems. As observed by Anderson (1990, p. 1575), "Examining hunger problems in the United States in terms of food security may allow both researchers and policymakers to confront this issue on a more objective basis."

While hunger has been an issue of public interest for a long time, the concept of household food insecurity emerged in the United States as a hunger-related concept relatively recently. Internationally (especially in the poorer countries and regions), the term "food insecurity" has been in use for some time to describe the inadequacy of national or regional food supplies over time. More recently, it has been expanded to include lack of food access at the household and individual levels (LeBlanc et al., 2005; Habicht, Pelto, Frongillo, and Rose, 2004). The LSRO conceptual definitions provided a basis for the USDA/DHHS initiative for developing operational definitions of food insecurity and hunger appropriate for use in large national population surveys.

The National Nutrition Monitoring and Related Research Act

Also in 1990, Congress enacted the National Nutrition Monitoring and Related Research Act (NNMRR) (Public Law 101-445). Section 103 of the act required the secretaries of the Department of Agriculture and the Department of Health and Human Services, with the advice of a board, to prepare and implement a ten-year comprehensive plan to assess the dietary and nutritional status of the U.S. population. Task V-C-2.4 in the plan specified (*Federal Register*, 1993, 58:32 752–806):

> Recommend a standardized mechanism and instrument(s) for defining and obtaining data on the prevalence of "food insecurity" or "food insufficiency" in the U.S. and methodologies that can be used across the NNMRR Program and at state and local levels.

A National Nutrition Monitoring Advisory Council was established on January 25, 1991, by Executive Order of the President as required by the NNMRR act (sect. 201 (a)(1)). The purpose of the advisory council was to provide scientific and technical advice on the development and implementation of the coordinated National Nutrition Monitoring and Related Research Program and the ten-year comprehensive plan and to serve in an advisory capacity to the secretaries of health and human services and agriculture.

Implementation of the Ten-Year Plan

Beginning in 1992, USDA staff systematically reviewed the existing research literature on the definition and measurement of food insecurity and hunger and on the practical problems of developing a survey instrument for use in sample surveys at the national, state, and local levels. This was the first step toward carrying out the responsibilities under the ten-year plan.

The Federal Food Security Measurement Project

Responding to the legislative requirements and to seek advice from a large group of experts, in 1992 USDA and DHHS brought together representatives from several federal agencies, academic researchers, private organizations, and other stakeholders to form the Federal Food Security Measurement Project. This interagency group developed over the course of several years the food security instrument, a set of food security scales that combine information from sets of questions in the instrument, and classification rules for characterizing the food security status of each household surveyed.

First National Conference on Food Security Measurement and Research

In January 1994, USDA and DHHS sponsored the First National Conference on Food Security Measurement and Research, which brought together a large group of experts from government, academia, and elsewhere who had worked in the area of identifying and measuring hunger and other aspects of food insecurity. This conference focused on issues of measurement and related research. The USDA and DHHS interest in measurement of food insecurity and hunger was that the measures developed be straightforward and relevant to public policy and policy makers and that they be scaled measures to reflect the variation in the level of severity of the condition observed. The purpose and objectives of the conference were to (U.S. Department of Agriculture and U.S. Department of Health and Human Services, 1994, p. iv):

- review the existing state of the art in operationalizing and measuring the dimensions of hunger and food insecurity in American households;
- clarify and seek consistency in the terminology employed in discussing resource-constrained hunger and food insecurity;
- explore the extent of consensus that has developed in the scholarly and research communities on the technical means of identifying and measuring resource-constrained food insecurity and hunger;
- obtain advice on the next steps needed to create a state-of-the-art survey instrument and database from which national prevalence measurement of food insecurity and hunger can be made; and
- consider some of the implications for research that would result from the availability of a standardized, annual national data set for the measurement of household-level hunger and food insecurity.

The conference identified the initial consensus among the participants on the appropriate conceptual basis for a national measure of food insecurity and on the technical means and feasibility of measuring hunger and food insecurity. The participants decided that food insecurity was the most important concept to measure, but some in the group held that hunger should be part of the measurement project as a device for advocacy (Habicht et al., 2004). The conference also resulted in a working agreement on several key issues, previously unresolved, as to the best measurement approach for implementation of a measure in national data collection, and the optimal content and form of a food security survey instrument for application at the national level. Conference participants decided:

1. to limit the measure to clearly poverty-linked or resource-constrained food insecurity and hunger and not attempt to measure hunger resulting from reasons other than resource constraints;
2. to limit operational definitions and measurement approach to those aspects of food security that can be captured in household-level surveys;
3. to focus on the behavioral and experiential dimensions of food insecurity and hunger (which were seen as the major gap in existing information and an essential component for policy makers);
4. to estimate the prevalence of food insecurity and hunger from the resulting data by scaling items into a single measure across all observed levels of severity of the phenomenon being measured if feasible; and
5. to develop a standard set of prevalence estimates at several designated levels of severity for consistent application and comparison across data sets from year to year. Participants further noted that agencies involved in collecting individual-level data might develop complementary approaches for measuring food insecurity at the specific individual level, whereas issues of community food security would require a different data collection strategy and orientation, outside the scope of the present effort (Hamilton et al., 1997a).

As a follow-up to the conference, USDA held additional meetings with the interagency working group, interested conference participants, and the Census Bureau staff to further explore, develop, and expand on the themes articulated in the conference. In order to follow up on the technical issues surfaced during the conference, USDA also commissioned additional analytical work based on two independent data sets of comprehensive hunger and food insecurity indicator items: the data set developed by the research group at Cornell University Division of Nutritional Sciences and the data set developed by the CCHIP (U.S. Department of Agriculture and U.S. Department of Health and Human Services, 1994).

FOOD SECURITY SUPPLEMENT TO THE CURRENT POPULATION SURVEY

In February 1994, USDA entered into an interagency agreement with the Census Bureau to develop, test, analyze, and refine a food security questionnaire as a supplement to the April 1995 Current Population Survey (CPS). The draft version of the questionnaire from the research conference was revised after deliberations and extensive cognitive testing and review by an expert team from the Center for Survey Methods Research (CSMR) and the Current Population Survey Branch of the Census Bureau. Technical direction of these extensive survey method refinements was provided by Eleanor Singer and the CSMR. The revised questionnaire was field-tested in April 1994. The results of the field test were analyzed by the fall of 1994, and further revisions were made to the questionnaire based on the recommendations flowing from the analysis of the pretest results (Hess, Singer, and Ciochetto, 1996). The questionnaire was fielded as a supplement to the CPS of April 1995 (U.S. Department of Agriculture and U.S. Department of Health and Human Services, 1994; Andrews, Bickel, and Carlson, 1998).

The CPS is a representative national sample survey of about 60,000 households that are surveyed monthly by the U.S. Census Bureau for the U.S. Department of Labor. It is a probability sample based on a stratified sampling design of the civilian, noninstitutionalized population. The CPS is the primary source of information on labor force characteristics of the United States. Various federal agencies sponsor collection of specialized supplementary data by the CPS following the labor force interview.

The Food Security Supplement (FSS) has been added to the CPS every year since 1995 (the supplement appears in Appendix A). It was repeated in September 1996, April 1997, August 1998, April 1999, September 2000, April 2001, December 2001, and in December of subsequent years. During 1996 and 1997, USDA made minor modifications to the questionnaire format and screening procedures. More substantial revisions in screening and format were introduced in 1998 to reduce respondent burden and improve data quality. The full FSS to the CPS includes more than 70 questions (including 2 follow-ups of about 15 questions) regarding expenditures for food, various aspects of food spending behavior and experiences during the 30 days and 12 months prior to the interview, use of federal and community food programs, food sufficiency and food security, and coping strategies.

Within the FSS is the Household Food Security Survey Module (HFSSM)—a set of 10 questions for households with no children and 18 questions for households with children—that is used to calculate the household food security scale and then to estimate the prevalence of food insecurity (see Box 2-1). This set of questions appears in the section on food sufficiency and security in the FSS.

The questions in the HFSSM have remained essentially unchanged over the years. These questions elicit information on whether the household experienced difficulty in meeting basic food needs due to a lack of resources. The severity of the food access problems covered by the food insecurity questions ranges from "worry about running out of food" to "children ever not eating for a whole day." The questions specify that any behavior or condition must be due to a lack of economic or other resources to obtain food, so the scale is not affected by hunger due to voluntary dieting, fasting, or being too busy to eat or other similar reasons. In an effort to keep respondent burden and annoyance low and at the same time not miss very many food-insecure households, the food insecurity questions and questions about ways of augmenting inadequate food resources are asked only of households with incomes below 185 percent of the federal poverty line and of households with incomes above that line if they give any indication of food access problems on either of the two screener questions (described in Chapter 4). Households with annual incomes above 185 percent of the poverty line and who give no indication of food access problems on these preliminary screener questions (one at the start of the section on food assistance program and participation, and the second at the start of the section on food sufficiency and security) are presumed to be food secure and are not asked subsequent questions. USDA analysis has shown that, given the screeners, only a very small number of food-insecure households are missed.

Research Activities on the Food Security Supplement

USDA undertook a considerable amount of research after fielding the supplement in 1995. As a condition of the Terms of Clearance for the April 1995 Food Security Supplement to the CPS, the Office of Management and Budget requested that the Census Bureau's Center for Survey Methods Research conduct an evaluation of the supplement questionnaire. Hess and colleagues (1996) conducted the evaluation, compared the results to those obtained during a pretest of the questionnaire conducted in August 1994, and provided recommendations for revising the questionnaire based on their evaluation.

Initial analysis of the data from the 1995 FSS was undertaken by Abt Associates, Inc., through a contract with USDA, and in consultation with the interagency working group on food security measurement and other key researchers involved in developing the questionnaire. The analysis focused on developing and implementing a measure of the severity of food insecurity. The aim was to help develop and assess a scale based on the FSS

Box 2-1
Questions Used to Assess the Food Security of Households in the CPS Food Security Supplement

1. "I/We worried whether my/our food would run out before I/we got money to buy more." Was that often, sometimes, or never true for you/ your household in the last 12 months?

2. "The food that we bought just didn't last and we didn't have money to get more." Was that often, sometimes, or never true for you in the last 12 months?

3. "We couldn't afford to eat balanced meals." Was that often, sometimes, or never true for you in the last 12 months?

4. In the last 12 months, did you or other adults in the household ever cut the size of your meals or skip meals because there wasn't enough money for food? (Yes/No)

5. (If yes to Question 4) How often did this happen—almost every month, some months but not every month, or in only 1 or 2 months?

6. In the last 12 months, did you ever eat less than you felt you should because there wasn't enough money for food? (Yes/No)

7. In the last 12 months, were you ever hungry, but didn't eat, because you couldn't afford enough food? (Yes/No)

8. In the last 12 months, did you lose weight because you didn't have enough money for food? (Yes/No)

9. In the last 12 months did you or other adults in your household ever not eat for a whole day because there wasn't enough money for food? (Yes/No)

---

that measures the severity of deprivation in basic food needs as experienced by U.S. households, to consider technical issues that arose in the development of the scale, and to produce a measurement scale for the severity of food insecurity (for a detailed description of this analysis, see Hamilton et al., 1997a, 1997b).

Following the collection of the 1996 and 1997 CPS food security data, USDA contracted with Mathematica Policy Research, Inc., to use multiple years of data from the FSS to consider empirical issues that had arisen, such

10. (If yes to Question 9) How often did this happen—almost every month, some months but not every month, or in only 1 or 2 months?

(Questions 11-18 are asked only if the household included children age 0-18)

11. "We relied on only a few kinds of low-cost food to feed our children because we were running out of money to buy food." Was that often, sometimes, or never true for you in the last 12 months?

12. "We couldn't feed our children a balanced meal because we couldn't afford that."
Was that often, sometimes, or never true for you in the last 12 months?

13. "The children were not eating enough because we just couldn't afford enough food."
Was that often, sometimes, or never true for you in the last 12 months?

14. In the last 12 months, did you ever cut the size of any of the children's meals because there wasn't enough money for food? (Yes/No)

15. In the last 12 months, were the children ever hungry but you just couldn't afford more food? (Yes/No)

16. In the last 12 months, did any of the children ever skip a meal because there wasn't enough money for food? (Yes/No)

17. (If yes to Question 16) How often did this happen—almost every month, some months but not every month, or in only 1 or 2 months?

18. In the last 12 months, did any of the children ever not eat for a whole day because there wasn't enough money for food? (Yes/No)

SOURCE: Nord, Andrews, and Carlson (2005b)

as the stability of the measurement scale over time, temporal adjustments to the categories for classifying severity of food security, the appropriate methods for assessing changes in the prevalence of food insecurity in the United States, screening issues, and imputation for missing data, among others (see Ohls, Radbill, and Schirm, 2001). The study focused on issues that arise in the development of a stable and consistent ongoing social indicator and issues that are critical when prevalence is measured on a routine basis and changes in prevalence are closely monitored by policy makers. The findings of this study established the stability of the food security scale over the 1995–1997 period. At the same time, the authors recognized some of the limitations of the model, particularly the process used to translate the continuous food security scale into food security categories, which is more judgmental than the process of deriving continuous scale scores. On the basis of their analyses, the authors suggested several directions for future research.

Under contract with USDA, IQ Solutions, Inc., assessed methodological issues and provided guidance with a specific focus on the 1998 and 1999 data and on measuring the food security of children in the first five years of CPS data collection (see Cohen et al., 2002). The focus of this study was to explore key technical issues related to the FSS, including techniques for the estimation of standard errors, the effect of alternating survey periods between spring and fall for the 1995–1999 CPS Supplement, and the effect of using different item response theory modeling approaches and software to create the food security scale.

Finally, USDA entered into a cooperative agreement with a group of statisticians and economists at Iowa State University (1997–2003) to conduct research designed to strengthen and improve the measurement of food security and hunger. Their research considered various statistical issues in measuring food insecurity and hunger and specifically the statistical properties of the Rasch model, which is used to scale responses to the FSS (see Opsomer, Jensen, Nusser, Dringel, and Amemyia, 2002; Froelich, 2002; Opsomer, Jensen, and Pan, 2003).

The study had two main purposes: to evaluate the robustness of the approach used for the measurement of food insecurity and to measure the effect of household-level variables on measured food insecurity. Both purposes were addressed by fitting the food security data with a class of models that generalizes the Rasch model and comparing the estimates obtained from the different models on several years of CPS data.

An initial product of this work was a review of the statistical properties of the Rasch model used by USDA to obtain estimates of the prevalence and severity of poverty-linked food insecurity and hunger in the United States (Opsomer et al., 2002). The authors point to several issues intended as a basis for discussion concerning future directions for research and plans of work, raising questions about the current approach for estimating the severity and prevalence of food insecurity and hunger as implemented by USDA. They identified a number of important issues in the estimation of food insecurity in the United States, including the use of a one-parameter logistic item response model (also referred to as a Rasch model); question construction; the impact of changes introduced in the 1998 CPS questionnaire, especially the introduction of new skip patterns that need to be evaluated; possible uses for the responses to the 30-day items; and measuring food insecurity in certain subpopulations. Opsomer and colleagues (2002) concluded that "food insecurity is a multidimensional concept, experienced differently by different household types and population groups. While an overall measure of food insecurity, valid for the whole U.S. population, would be desirable, it is likely that such a measure would underestimate hunger and food insecurity for certain subgroups, especially for children and elderly adults. . . . Food insecurity is a complex issue that may not be fully captured by a one-dimensional item response model, especially as it will be used to track food insecurity over time, across different surveys, and for different subpopulations" (p. 25).

Opsomer and colleagues (2003) took a closer look at the measurement of food insecurity and the effect of household variables on measured food insecurity. Using data from the 1995, 1997, and 1999 food security module of the CPS, they evaluated the effects of demographic and survey-specific variables on the food insecurity/hunger scale using a generalized linear model with mixed effects. The Rasch model used by USDA assumes that the food insecurity questions are interpreted in the same manner by all households interviewed. If this assumption does not hold, the estimated question scores and the estimates derived from them are potentially biased. The generalized linear mixed model used by the authors makes it possible to incorporate household variables as well as interactions between these variables and the question scores. Having these types of variables explicitly in the model provides answers to such questions as: Are certain demographic groups more likely to be food insecure? Are survey mode effects present in the survey? Are certain questions understood differently by certain demographic groups? Opsomer and colleagues (2003) generally validated the model used by USDA across time periods; however, there was some evidence that interpretation of questions may vary among demographic groups, a conclusion also reached by other researchers. The authors recommended further research to understand the robustness of the method across different subgroups and in the context of alternate survey modes and experiences of hunger.

Froelich (2002) examined the aspect of dimensionality in the USDA food insecurity measure using other analysis from the field of item response theory. The study focused on the dimensionality of the food security index for households with children. The paper introduces the item response theory definition of dimensionality, describes two different exploratory dimensionality analyses, and reports on the results of these analyses. The analyses provided evidence of the presence of two dimensions in the USDA food insecurity scale for households with children. The items in the first dimension measure food insecurity and hunger of the household adults, and the items in the second dimension measure the food insecurity of the children in the household. Froelich concluded that further research is needed to find the amount of the potential bias of the item severity estimates used to construct the scale and to determine a method of dealing with the multidimensionality of food insecurity and hunger present in households with children.

Second Conference on Food Security Measurement and Research

In 1999, USDA and DHHS hosted the Second Conference on Food Security Measurement and Research. This conference focused on priorities for a future research agenda. Efforts were made to ensure a wide range of perspectives and to solicit critical review of the standard measure and prior research. The papers and proceedings of the conference were published by USDA (Andrews and Prell, 2001a, 2001b).

The Interagency Working Group on Food Security Measurement met after the conference to review the proceedings and identified a set of research priorities that flowed from the conference. The major themes of highest priority are shown in Box 2-2.

USES OF THE HOUSEHOLD FOOD SECURITY SURVEY MODULE IN OTHER SURVEYS

The federal food security measurement project has developed standardized questionnaires and methods for producing household summary measures of food security status. These modules are, or have been, used at times with modifications in several national surveys and in a growing number of state, local, and regional studies. Also, there are several instances of their adaptation for use in other countries. Examples of some of these uses are briefly summarized below.

Surveys in the United States

The food security module, or a modification of it, has been incorporated at some time in national surveys by the various agencies participating in the Federal Food Security Measurement Project. Some of these surveys are shown below.

- The National Health and Nutrition Examination Survey (NHANES IV), conducted by NCHS, has included the 18-item HFSSM in the family interview part of the household interview since 1999. Individually referenced food security questions (7 for participants age 16 and older, 6 for participants younger than age 16) were added to the postdietary recall component of the examination section and were released with the 2001–2002 food security data release. NHANES continues to collect both the household- and individual-level food security data.

Box 2-2
Priority Research Agenda

Research Priorities: Measurement

- Development and testing of individual (as opposed to household) scales for measurement of prevalence of food insecurity among adults and children;
- Improvements in the measurement and understanding of the dynamics of food insecurity, such as frequency and duration of episodes;
- Developing better questions and strategies for asking about nutritional quality (alternative to balanced meal questions);
- Assessment of the effects of the questionnaire structure, item sequencing, and survey context on response patterns and measured food security levels; and
- Determination of research situations appropriate for implementation of abbreviated household food security scales and/or scales with different time frames such as monthly versus annual.

Research Priorities: Applications and Policy

- Focus on sampling and research on food insecurity and its consequences among high-risk groups with chronic health conditions, mental illness, and other biological vulnerability (especially among the homeless, elderly, and young children);
- Development of a research basis for linking community food security and household food insecurity;
- Better understanding of the context and determinants of food insecurity and hunger and their relationship to poverty, household resources, and time management; and
- Applications that assess and investigate the linkages between food insecurity measures, welfare reform, and measures of program performance.

Source: Andrews and Prell (2001a)

- The Division of Nutrition of the Centers for Disease Control and Prevention (CDC), NCHS, and USDA have worked together to develop subscales of the 18-item scale, such as a 6-item set, that could be used to measure food insecurity and hunger in state surveillance systems, such as the NCHS State and Local Area Integrated Telephone Survey and the CDC Pediatric Nutrition Surveillance System.

- The Census Bureau's Survey of Program Dynamics, fielded for five consecutive years beginning in 1998, included the 18-item food security module.
- The Survey of Income and Program Participation (SIPP) included a subset of six questions (but not the standard six-item set) in the adult well-being module once during each panel beginning in 1998 (see Chapter 6 for further description of SIPP).
- The Early Childhood Longitudinal Survey, Kindergarten Class of 1998–1999 (ECLS-K) and Birth Cohort of 2002 (ECLS-B), incorporated the HFSSM. ECLS-K and ECLS-B are longitudinal surveys of the U.S. Department of Education's National Center for Education Statistics. ECLS-K follows a nationally representative sample of about 22,000 children from kindergarten to fifth grade (and likely beyond); ECLS-B follows a nationally representative sample of about 10,000 newborn children for their first several years.
- The University of Michigan Panel Survey of Income Dynamics included the household food security survey module in a special supplement on women and children in 1997 and in the full sample beginning in 1999.
- The Behavioral Risk Factor Surveillance System included questions on food sufficiency in the social context module. Eight states collected the information at some point between 1996 and 1999.

A number of state surveys used all or a part of the HFSSM:

- Oregon included the short six-item food security module in the Oregon Population Survey in 2000.
- The California Health Interview Survey included the short six-item food security module beginning in 2001.
- Welfare, Children and Families: A Three City Study was conducted in Boston, San Antonio, and Chicago. The study is a collaboration among several universities. It included limited food security and hunger questions, some of which appear to be taken directly from the Food Security Supplement.
- The Wisconsin Food Security Survey in the WIC population included the six-item limited interview. The survey was translated into Spanish, Hmong, and Russian.
- Iowa included the six-item food security module in a survey of WIC participants in English and Spanish beginning in 2003.

International Adaptations

Several countries have developed food security measures based on approaches similar to those of the United States. At times this is achieved with simple translation of the questions in the U.S. module into the local language. At other times to achieve acceptable results, the U.S. measure has required adaptation to settings that may be culturally, linguistically, and economically different from the United States; this may require additional research, including focus groups and cognitive testing of proposed questions and statistical analysis of survey data. Such adaptations have been used in low-income populations in Orissa, India; Kampala, Uganda; and Bangladesh (see Nord, Sathpathy, Raj, Webb, and Houser, 2002, for a detailed description). Still other efforts in Bangladesh and Burkina Faso have taken the approach that was used to develop the U.S. module by conducting ethnographic interviews on people's experience of food insecurity, developing questionnaire items, and testing them (Frongillo, Chowdhury, Ekström, and Naved, 2003; Frongillo and Nanama, 2003; Webb, Coates, and Houser, 2003).

The following are brief summaries of some of the more significant international adaptations of the U.S. food security measurement methods.[1] These are all nationally representative surveys. Pilot surveys and surveys of targeted populations have been conducted in many other countries.

- Israel conducted a national-level food security assessment as part of a national health and nutrition survey in 2003. The data were analyzed and reported by the JDC-Brookdale Institute (an academic research institution) in partnership with the Ministry of Health, the National Insurance Institute, the Ministry of Social Affairs, and the Forum to Address Food Insecurity and Poverty in Israel.
- Brazil—The government of Brazil conducted a nationally representative food security assessment in 2004 in connection with a national health and nutrition survey. The data have not yet been analyzed. The measurement project in Brazil has very high political visibility because the current president of Brazil ran on a platform of "zero hunger." Results of one of the pilot surveys on which the final survey module and methods were based were published in the *Journal of Nutrition* (Pérez-Escamilla et al., 2004).

---

[1]Email communication with Mark Nord, Economic Research Service, USDA, on April 29, 2005.

- Colombia—Collection of nationally representative survey data on food security in Colombia is currently in progress. A report based on data collected in one province to pilot and finalize the methodology is available (in Spanish only), "Perfil Alimentario y Nutricional de los Hogares: Analisis Comparativeo entre las Regiones de Antioquia."
- Serbia, Yugoslavia—United Nations Food and Agriculture Organization (FAO) staff assessed food security in the Republic of Serbia (Yugoslavia) in 2002 through five coordinated household surveys. The Economic Research Service of USDA provided technical assistance on data assessment and analysis. The results were used in FAO but have not been published.
- Argentina—The World Bank conducted a food security assessment in Argentina in connection with a survey to assess the social impact of a recent economic crisis.
- Yemen—A nationally representative assessment of food security was conducted in Yemen in 2003. The results have not yet been officially reported.

In summary, the Food Security Supplement to the CPS is the cornerstone of the Federal Food Security Measurement Project, which began in 1992 to carry out a key task assigned by the Ten-Year Comprehensive Plan, namely, to develop a standard measure of food insecurity and hunger for the United States for use at the national, state, and local levels. A large body of literature has developed from research, both internal and external to USDA and DHHS, covering methodological topics related to the measurement of food security, and the measure has been adapted for use in several other countries. This research has prompted further refinements and modifications to the food security questionnaire, including among other things a shorter, six-item food security module and measure, separate adult and child food security measures, a revised 30-day measure, and the translation of the survey module into Spanish. The research has also raised questions that USDA should address.

3

Concepts and Definitions

T his chapter discusses the conceptual issues associated with the concepts and definitions of food insecurity and hunger and their applications for measurement in the monitoring of food insecurity in the United States. The chapter also discusses the labeling of the severity levels of food insecurity.

FOOD INSECURITY, HUNGER, MALNUTRITION, AND UNDERNOURISHMENT[1]

Food scarcity, with its dangers for survival and serious physical and psychological discomfort, has been part of human experience and human culture from the earliest inception of language and thought. Various concepts have emerged to describe aspects and consequences of food scarcity, although they are often ambiguous in meaning. For example, depending on usage and the user, the concept of hunger covers a spectrum from the short-term physical experience of discomfort to chronic food shortage to severe and life-threatening lack of food.

With the establishment of the modern science of nutrition, the concept of *malnutrition* as a condition brought about by insufficient intake of nutrients to meet biological requirements became a focal construct. Technically the prefix *mal* actually refers to both over- and underintake, but the typical

---

[1]This section is adapted from Habicht et al. (2004).

usage—and until recently the bulk of research on malnutrition—has been directed to understanding inadequate intakes of macro- and micronutrients. The measures of central concern are observed through analysis of biological tissues (e.g., serum), observation of well-established physical (e.g., anthropometric) and clinically observable consequences (e.g., blindness), and by inference from data on intake. For example, anthropometric status is commonly used to assess malnutrition of children under age 5 (de Onis, Blösner, Borghi, Frongillo, and Morris, 2004).

As malnutrition acquired a central role in scientific conceptualization, it was often mentioned jointly with the idea of hunger, to the point at which the two often became virtually synonymous. Nutritional scientists as well as social advocates therefore sought to describe the inequalities of access to adequate food and its consumption. One approach was to compare intakes of a nutrient for a given gender and life stage group with an established reference value, such as the Recommended Dietary Allowances (RDAs).

Some problems with using the RDA approach stem, in part, from its conceptual underpinnings. To cover the needs of nearly all of a group, the reference values were set at very high levels. Consequently, a proportion of the population may consume less than the RDAs but still have adequate nutrient intakes. Another problem is purely technical. It is difficult to use a single interview to assess usual nutrient intake in a biologically meaningful fashion. For instance, vitamin A intake varies considerably over time, and only the mean intake over a period of weeks is meaningful nutritionally, because vitamin A is stored and body reserves buffer the variability of intake. Further technical problems relate to the accuracy of reported intake and of the information used to translate food intake into nutrients. As a consequence of these problems, assessment of nutritional adequacy through interviews and analysis of the record in relation to the RDA is no longer considered appropriate (Institute of Medicine, 2000).

The United Nations Food and Agricultural Organization (FAO) took a different biologically based approach to define *undernourishment* as not ingesting enough food to meet energy needs. Operationally the FAO indicator is calculated from national food energy balance sheets. These balance sheets estimate the total energy available for human consumption nationally by adding total energy produced plus energy imported plus the change in stocks minus energy exported, energy wasted, and energy used for other than human consumption. FAO then creates a synthetic distribution of energy consumption for each country in which the mean is total energy available (from the balance sheets) and the variance is taken from another source, typically an estimate from a nationally representative household expenditure survey that accounts for energy exported and energy used for other than human consumption (Naiken, 2003). The resulting estimated distribution of undernourishment (i.e., food energy consumed) across countries is highly correlated with the distribution of food energy available for consumption obtained directly from the national food energy balance sheets when national population size is taken into account (Smith, 1998). Thus the two measurements, one from the energy balance sheets and one from the prevalence of undernourishment, are redundant. That is, the FAO method for estimating undernourishment measures only food energy availability, but not consumption of (or access to) food by households.

The discovery that people frequently did not have enough to eat according to accepted cultural norms created a conceptual crisis. Either the food problems of poor people were imaginary, or other concepts were needed to describe and measure them. An intuitively understandable construct was *hunger* defined as a physical pain. This word has typically and historically been used not only to refer to the physical sensation, but also to a feeling of weakness from not eating. As stated in the previous chapter, beginning in the 1960s, the word hunger began to take on a wider meaning. It was expanded to encompass issues of access to food and socioeconomic deprivation related to food. Perhaps because these expanded referents seemed less compatible with the intuitive meaning of hunger, other constructs were needed. It is in this context that the phrase *food insecurity* came into use in the United States. Internationally, food insecurity was already current. Originally, it was used to describe the instability of national or regional food supplies over time (Pelletier, Olson, and Frongillo, 2001; Rose, Basiotis, and Klein, 1995). It was then expanded to include a lack of secure provisions at the household and individual levels.

Figure 3-1 depicts the core concepts related to nutritional state that were established at the commencement of the U.S. national nutritional monitoring system (Anderson, 1990).

CONCEPT AND DEFINITION OF FOOD INSECURITY

As described in the previous chapter, the broad conceptual definitions of food security and insecurity developed by the expert panel convened in 1989 by the Life Sciences Research Office (LSRO) have served as the basis for the standardized operational definitions used for estimating food security in the United States. *Food security* according to the LSRO definition means access to enough food for an active, healthy life. It includes at a minimum (a) the ready availability of nutritionally adequate and safe foods and (b) an assured ability to acquire acceptable foods in socially acceptable ways (e.g., without resorting to emergency food supplies, scavenging, stealing, or other coping strategies). *Food insecurity* exists whenever the availability of nutritionally adequate and safe foods or the ability to acquire acceptable foods in socially acceptable ways is limited or uncertain.

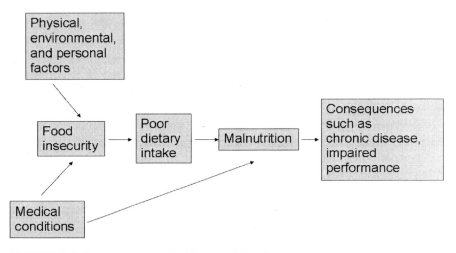

FIGURE 3-1 Core concepts related to nutritional state.

Food insecurity, as measured in the United States, refers to the social and economic problem of lack of food due to resource or other constraints, not voluntary fasting or dieting, or because of illness, or for other reasons. This definition, supported by the ethnographic research conducted by Radimer et al. (1992); Wolfe, Frongillo, and Valois (2003); Hamelin, Habicht, and Beaudry (1999); Hamelin, Beaudry, and Habicht (2002); Quandt and Rao (1999); Quandt, McDonald, Arcury, Bell, and Vitolins (2000); and Quandt, Arcury, McDonald, Bell, and Vitolins (2001), means that food insecurity is experienced when there is (1) uncertainty about future food availability and access, (2) insufficiency in the amount and kind of food required for a healthy lifestyle, or (3) the need to use socially unacceptable ways to acquire food (see Figure 3-2). Although lack of economic resources is the most common constraint, food insecurity can also be experienced when food is available and accessible but cannot be used because of physical or other constraints, such as limited physical functioning by elderly people or those with disabilities (Lee and Frongillo, 2001a, 2001b).

Some closely linked consequences of uncertainty, insufficiency, and social unacceptability are assumed to be part of the experience of food insecurity. Worry and anxiety typically result from uncertainty. Feelings of alienation and deprivation, distress, and adverse changes in family and social interactions also occur (Hamelin et al., 1999, 2002; Frongillo and Horan, 2004). As stated in the previous chapter, hunger and malnutrition are also potential, although not necessary, consequences of food insecurity. Management strategies that people use to prevent or respond to

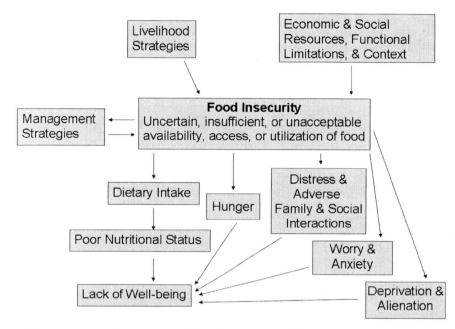

FIGURE 3-2 Food insecurity, and its determinants and consequences (adapted from Habicht et al., 2004).

the experience of food insecurity are conceptually different from food insecurity but are tied to it.

Food insecurity is measured as a household-level concept that refers to uncertain, insufficient, or unacceptable availability, access, or utilization of food. It is experienced along with some closely linked consequences of it. There is a strong rationale for measuring food insecurity at the household level. It is possible for individuals to be food secure in a food-insecure household, just as it is possible for individuals to not be poor in a poor household, depending on the intrahousehold allocation of resources. It means that we can measure and report the number of people who are in food-insecure households (with not all of them necessarily food insecure themselves). When a household contains one or more food-insecure persons, the household is considered food insecure.

Although food is a fundamental need in that each individual must have access to necessary nutrients to survive and to participate actively in society, food is only one of the needs that people must make efforts to meet. Households often make trade-offs among needs to ensure their long-term viability as units. Households manage the stocks and flows of assets and cash to meet basic needs, offset risk, ease shocks, and meet contingencies (Pelletier et al., 2001; Rose et al., 1995). For example, people in households may consume less food in the present to preserve assets and future ability to make their living, or people may forgo some food to be able to buy medication to treat illness (Wolfe et al., 2003). A full understanding of food insecurity requires incorporation of the time element—both in the sense of the periodicity of occurrence of various needs and events and in the sense of the frequency and duration of episodes (Maxwell and Frankenberger, 1992). *Frequency and duration are therefore important elements for the U.S. Department of Agriculture (USDA) to consider in the operational definition and measurement of household food insecurity and individual hunger.* (This issue is discussed further in Chapter 4.)

ADVERSE OUTCOMES OF FOOD INSECURITY

Research has shown that food insecurity is associated with adverse health and developmental outcomes in children and adults that are both nutritional and nonnutritional in nature.[2] Food insecurity is associated with higher prevalence of inadequate intake of key nutrients (Rose, Habicht, and Devaney, 1998; Casey, Szeto, Lansing, Bogle, and Weber, 2001; Lee and Frongillo, 2001a; Adams, Grummer-Strawn, and Chavez, 2003), risk of overweight in women and some girls (Olson, 1999; Alaimo, Olson, and Frongillo, 2001a; Laitinen, Power, and Javelin, 2001; Townsend, Peerson, Love, Achterberg, and Murphy, 2001), depressive symptoms in adolescents (Alaimo, Olson, and Frongillo, 2002), and academic and social developmental delays in children (Kleinman et al., 1998; Murphy et al., 1998; Alaimo et al., 2001b; Reid, 2001; Stormer and Harrison, 2003; Ashiabi, 2005). Data from a longitudinal study of welfare recipients show that household food insecurity is associated with poor physical and mental health of low-income black and white women (Siefert, Heflin, Corcoran, and Williams, 2004). Food insecurity is also associated with more behavioral problems (Olson, 1999; Shook Slack and Yoo, 2004), poorer school performance (Olson, 1999; Alaimo et al., 2001b; Dunifon and Kowaleski-Jones, 2003), and adverse health outcomes (Alaimo, Olson, Frongillo, and Briefel, 2001c; Cook et al., 2004; Weinreb et al., 2005) in children. Data from the Early Child Longitudinal Study-Kindergarten Class show that reporting at least one indicator of food insecurity was significantly associated with impaired learning in mathematics from fall to spring of the kindergarten year (Winicki and Jemison, 2003) and with impaired learning in reading from kindergarten to third grade (Jyoti, Frongillo, and Jones, 2005).

---

[2]The panel does not attempt to present a comprehensive review of all possible literature on the subject.

CONCEPT AND DEFINITION OF HUNGER

The conceptual definition of hunger adopted by the interagency group on the food security is: "The uneasy or painful sensation caused by a lack of food, the recurrent and involuntary lack of food. Hunger may produce malnutrition over time. . . . Hunger . . . is a potential, although not necessary, consequence of food insecurity" (Anderson, 1990, pp. 1575, 1576). This language does not provide a clear conceptual basis for what hunger should mean as part of the measurement of food insecurity. The first phrase "the uneasy or painful sensation caused by a lack of food" refers to a possible consequence of food insecurity, as discussed above. The second phrase "the recurrent and involuntary lack of access to food" refers to the whole problem of food insecurity, the social and economic problem of lack of food as defined above.

Holben (2005)[3] has enumerated a large number of definitions of hunger from various sources. Taken together, these definitions fall into four groups regarding the concept of hunger: (1) a motivational drive, need, or craving for food; (2) an uneasy sensation felt when one has not eaten for some time; (3) discomfort, illness, weakness, or pain caused by a prolonged, involuntary lack of food; and (4) the prolonged, involuntary lack of food itself. The first and second of these are not the interest of the household food security survey because they refer to a natural phenomenon that all humans experience on a regular basis. The fourth is also not a useful definition or concept of hunger because it refers to the problem of food insecurity itself. The third provides a starting point for consideration as to what is intended for the Household Food Security Survey Module (HFSSM). It refers to the *consequence* of food insecurity that, because of a prolonged, involuntary lack of food due to lack of economic resources, results in discomfort, illness, weakness, or pain that goes beyond the usual uneasy sensation.

Available evidence from ethnographic work affirms that this definition of hunger is well understood and is reported in similar terms in the United States (Radimer et al., 1992; Wolfe, Frongillo, and Valois, 2003) and Québec (Hamelin, Beaudry, and Habicht, 2002). There is consensus in U.S. society, supported by this empirical research, that an individual's report that he or she has experienced hunger because of lack of food provides a straightforward indication that the individual has, indeed, experienced hunger in the sense of the third definition (i.e., discomfort, illness, weakness, or

---

[3]This information is drawn from a background paper prepared for the panel by Holben (2005).

pain caused by a prolonged, involuntary lack of food). But unlike food insecurity, which is a household-level concept, hunger is an individual-level concept. For purposes of the HFSSM included in the Food Security Supplement to the CPS, the term "hunger" should refer to a potential consequence of food insecurity that, because of prolonged, involuntary lack of food, results in discomfort, illness, weakness, or pain that goes beyond the usual uneasy sensation. Two questions therefore arise. First, can the experience of severe food insecurity with hunger by households be measured and its prevalence estimated? Second, can the experience of hunger by individuals be measured and its prevalence estimated?

The HFSSM is measuring food insecurity at the level of the household; it is not measuring hunger at the individual level. The scale does not give special weight to the hunger questions. The HFSSM does include items that are related to being hungry among food-insecure households. The ethnographic and quantitative evidence discussed earlier has shown that the HFSSM items on hunger are probably appropriate in the food insecurity scale, but these items contribute to the measurement of household food insecurity and not specifically to the measurement of hunger at the individual level.

For the purposes of measuring and estimating the prevalence of hunger among individuals in the population, something that the HFSSM does not do, some of these same items might be used in a measure of hunger among individuals, but it would require a measurement process that is based on the conceptual definition of the condition, as well as a battery of items designed to measure it and a reoriented sampling design that includes the individual as the unit of analysis. This work could be based on the information from such sources as up-to-date ethnographic studies of low-income populations, results of experiments and analysis of surveys, analysis of public opinion and perspectives of user groups, expert assessment, and other relevant information.

The panel therefore concludes that hunger is a concept distinct from food insecurity, which is an indicator and possible consequence of food insecurity, that can be useful in characterizing severity of food insecurity. Hunger itself is an important concept, but it should be measured at the individual level distinct from, but in the context of, food insecurity.

To summarize, the panel's conclusion is based on the fact that, although a strong theoretical and research base exists for the conceptualization and measurement of food insecurity, we do not have a correspondingly strong base for either the conceptualization of hunger or its measurement. That is, there is now ample theoretical, conceptual, ethnographic, and quantitative work done to justify the measurement of the experience of food insecurity using a questionnaire. For the measurement of the experience of hunger to be equally credible, there needs to be a stronger base than we currently have in developing clear concepts for how we should think about hunger and in tested means to accurately elicit information from survey respondents about whether they have experienced hunger.

> Recommendation 3-1: USDA should continue to measure and monitor food insecurity regularly in a household survey. Given that hunger is a separate concept from food insecurity, USDA should undertake a program to measure hunger, which is an important potential consequence of food insecurity.

> Recommendation 3-2: To measure hunger, which is an individual and not a household construct, USDA should develop measures for individuals on the basis of a structured research program, and develop and implement a modified or new data gathering mechanism. The first step should be to develop an operationally feasible concept and definition of hunger.

> Recommendation 3-3: USDA should examine in its research program ways to measure other potential, closely linked consequences of food insecurity, in addition to hunger, such as feelings of deprivation and alienation, distress, and adverse family and social interaction.

It took a lot of discussion and conferences for the Food Security Measurement Project to reach a working agreement on the operational definition of food security and insecurity. Hunger is a complex concept, and it should be well thought through to ensure agreement among the key users and then to develop and test the appropriate questions and to identify the survey mechanism and sample design for collecting the needed data. Such an effort will take time.

APPLICATION OF THE CONCEPTS AND DEFINITIONS FOR MEASUREMENT

The broad conceptual definition of household food insecurity includes more elements than are included in the current USDA measure of food insecurity. The current measure of prevalence of household food insecurity obtained through the HFSSM focuses on the uncertainty and insufficiency of food availability and access that are limited by resource constraints, and the worry or anxiety and hunger that may result from it. It does not include questions on nutritional adequacy, safety, or social unacceptability of food access, concepts that are part of the broad conceptual definition.

It also does not include questions on use that may be particularly applicable to elderly people and those with disabilities. The HFSSM covers the core ideas of food being available and accessible but not the ability to be used; measurement of food insecurity is tied to economic constraints but not physical constraints that might affect use of food. Wolfe and colleagues (2003) point out that although economic constraint is a major cause of food insecurity, elderly people sometimes have enough money for food but are not able to access it because of transportation or functional limitations, or they are not able to use the food, that is, not able to prepare or eat available food. However, there is currently no epidemiological evidence demonstrating that incorporating items about the ability to use food will alter the prevalence estimates of food insecurity. Furthermore, there is no evidence to suggest that the original decision made when the measure was developed to focus on food insecurity that arises in the context of economic constraints—deemed to be the food insecurity that is policy-relevent—should be altered.

Furthermore, the HFSSM does not attempt to measure management strategies, although questions are asked in the Food Security Supplement (FSS) that do assess management (i.e., augmentation) strategies, such as getting emergency food from a food pantry, eating meals at a soup kitchen, and borrowing money to buy food. "These coping behavior items were tested for inclusion in the food security scale. However, they were found not to meet the statistical test criteria for inclusion with the measurement scale, even though they correlate closely with the scale. Very few households use these copying behaviors that are not also identified as food insecure by the scaled measure" (Bickel, Nord, Price, Hamilton, and Cook, 2000, p. 43).

As stated in the previous chapter, one of the requirements of the National Nutrition Monitoring and Related Research Act is to recommend a standardized mechanism and instrument for defining and obtaining data on prevalence of food insecurity or food insufficiency at the national and state levels. The Food Security Measurement Project working group reached agreement during the 1994 conference to limit the operational definitions and measurement to only those aspects of food security that can be captured in household-level surveys and to further limit the measure to lack of economic resources to obtain food. The definition does not include the supply of food or nutrition. These additional aspects would require developing measures and fielding separate surveys to measure them. The food supply in the United States is generally regarded as safe relative to some other countries. Nutritional adequacy is already assessed by other elements of the nutrition monitoring system, in particular the continuing National Health and Nutrition Examination Survey.

The panel therefore concludes that it is neither required nor necessarily appropriate for USDA to attempt to measure all elements of the conceptual definition of food insecurity as part of the HFSSM.

LABELS OF FOOD INSECURITY

Since food insecurity is conceptualized by USDA as a continuous scale score, attaching labels to various levels of the score to communicate the results in a simpler manner is a natural and common presentation device. As discussed in Chapter 5, it is a common practice to present estimates from scale scores by identifying cut points on the scale and characterizing the units between the cut points by a descriptive label. For example, a person scoring above a cut point might be considered to exhibit proficiency in a specific skill in a certification test, or cut points might be assigned that classify students' performance in mathematics in an educational assessment (e.g., see Adams and Wu, 2002). One goal of the cut points is to form a classification scheme that categorizes individuals (e.g., in a certification test). A second goal is to describe the distribution of the number of persons or households in each category of the classification scheme over the population (e.g., in the National Assessment of Educational Progress, the children assessed are not labeled individually, but the proportion of children at each level is estimated). Methods for establishing cut points to define a classification scheme are discussed in Chapter 5.

The labels for the categories associated with the units between the cut points are themselves very important because they are the vocabulary used in the discussion of the scores. In fact, the labels are the primary way of identifying the outcomes for many audiences for whom the score itself may be of less interest. For example, journalists, the public, and secondary users of the data typically discuss estimates almost exclusively in terms of the labels. An analogy is the relationship between income, a continuous variable similar to a scale score, and poverty, which is a label, based on a cut point of income (and family size). Many users are almost exclusively interested in the number and characteristics of persons below the "poverty" level.

Because of their importance, the labels should be consistent with the data collected and should communicate a common understanding of what is being measured. The recent report issued by the Committee on Performance Levels for Adult Literacy discusses these as important criteria for labels (National Research Council, 2005). In that report, the committee explicitly recommended not using "proficiency" in any of the labels for the 2003 National Assessment of Adult Literacy (NAAL) assessments, because the test was not intended to measure proficiency, nor can it be revised in a post hoc manner to do so.[4] The committee concluded that labeling adults as proficient based on the NAAL assessments would be inconsistent with the data collected and the common understanding of the term "proficient." This recommendation points out the importance of the labels and the requirement that they be appropriate.

The labeling used for the classification of food insecurity is at the heart of the criticism of the current measurement system. In particular, the category "food insecurity with hunger" has come under scrutiny because of disagreement over whether hunger is actually measured. For the current discussion, we ignore the distinction between households with and without children and discuss only the labels for households without children—food secure, food insecure, and food insecure with hunger. A key criticism of the current system is that a household may be labeled "food insecure with hunger" even though the household respondent does not explicitly affirm hunger in the interview. This criticism is discussed earlier in this chapter and is further considered in Chapter 5.

The rationale for including hunger in the label for the scale is understandable. Hunger is a politically sensitive and evocative label that conjures images of severe food deprivation, and the HFSSM includes some items that are specifically related to hunger. As discussed here and in Chapter 2, however, the measurement of food insecurity rather than hunger has been the primary focus of the HFSSM since its inception.

A particular concern, which has been raised and discussed earlier in this chapter, is that hunger as experienced by an individual and hunger as experienced by persons in a household may differ. As an indication of the severity of food insecurity, the HFSSM asks the household respondent if in the past 12 months she or he has experienced being hungry because of lack of food due to resource constraints. This is not the same as evaluating all individuals in the household in a survey as to whether or not they have experienced hunger.

A second concern is that, in some households with severe food insecurity, none of the household members may be hungry, while in other households some members will be hungry and some will not. Food insecurity has other consequences, many of which may have effects that are serious and long-lasting.

To illustrate the panel's concerns, consider the USDA report containing the basic estimates from the 2004 CPS supplement on food security (Nord et al., 2005b). While this report carefully explains the concepts and issues associated with food insecurity and how it is measured, a section (page 7) is

---

[4]The report recommends that adults assessed in the 2003 NAAL be classified into five performance levels using the following labels: nonliterate in English, below basic literacy, basic literacy, intermediate literacy, and advanced literacy.

titled "How often were people hungry in households that were food insecure with hunger." The point of the section is that food insecurity is typically episodic rather than chronic in the United States, and most of the details in the section clearly state that the estimates refer to the classification of households, not hunger in individuals. The title suggests, however, that the survey addresses the question of how often people are hungry, even though the survey does not produce valid estimates of how often people are hungry. Despite careful wording within the section, the title is misleading.

Another example is the USDA report on measuring food security for children (Nord and Bickel, 2002). Again, this report carefully explains the concepts and issues associated with food security and how it is measured and addresses a serious concern about the effect of the classification scheme for a subgroup of great importance, children. Despite the authors' clear understanding of the key issues and care in presentation, the report discusses the prevalence of hunger among children, even in the abstract. The labeling of the most severe range as food insecure with hunger adds confusion to the reporting.

These two reports are actually among the best in terms of expression of the concepts and using the data appropriately, and yet even they could be easily misinterpreted. Other examples could be cited from documents that are less carefully worded and distort the data from the survey. USDA needs to be more careful in its reports to properly communicate that this third category refers to households with severe food insecurity in which the respondent has either missed meals or was hungry because there wasn't enough money for food at some time during the year.

With a label that includes the word "hunger," challenges in communicating an appropriate understanding of food insecurity are inevitable. The challenges are intrinsic, in that the label conveys the idea that severe food insecurity is synonymous with hunger, while this is not necessarily the case. Modifications of the label that still include the word "hunger" do not eliminate the potential for misleading many users, especially the public. Measuring prevalence of household food insecurity with the respondent experiencing hunger is not the same as measuring the prevalence of hunger experienced by individuals. The latter will require a separate research and development process to be implemented in an individual respondent–based survey as opposed to a household respondent–based one.

Alternate labels that may be less problematic could be used. The panel urges USDA to consider alternate labels that may be better and to develop short and appropriate descriptions of the types of households that fall within the cut points associated with the labels. The report of the Committee on Performance Levels for Adult Literacy, referred to above, has given guidance on these types of descriptions that are equally valid for the food insecurity scale.

Recommendation 3-4: USDA should examine alternate labels to convey the severity of food insecurity without the problems inherent in the current labels. Furthermore, USDA should explicitly state in its annual reports that the data presented in the report are estimates of prevalence of household food insecurity and not prevalence of hunger among individuals.

4

Survey Measurement of
Food Insecurity and Hunger

This chapter reviews the current measurement of food insecurity, examines the questions used to measure food insecurity and food insecurity with hunger, and identifies several problems in the design of the Household Food Security Survey Module (HFSSM) that should be addressed.

CURRENT APPROACH TO MEASUREMENT

Design of the Current Population Survey

As summarized in Chapter 2, the Food Security Supplement (FSS) is conducted as a supplement to the Current Population Survey (CPS). The FSS was first included in the CPS in April 1995 and has since been administered every year—September 1996, April 1997, August 1998, April 1999, September 2000, April 2001, December 2001, December 2002, December 2003, and December 2004.

The CPS is a representative national sample survey of the civilian, non-institutional population, conducted monthly by the U.S. Bureau of the Census for the U.S. Department of Labor. Approximately 60,000 households are interviewed each month with data collected on the labor force participation status of approximately 119,000 individuals. The CPS uses a rotating panel of households that are interviewed for four consecutive months, not interviewed for the next eight months, and then interviewed for four additional months, for a total of eight interviews. For any given month, the data represent interviews collected from eight rotation groups, that is, one-eighth of the sample is included for the first time, one-eighth is in the sample for the second time, etc. The sample unit is the housing unit; if residents move over the course of the 15 months, the remaining CPS interviews are conducted with the new members of the housing unit. For this reason, the information collected may not represent panel data per se. The initial interview and the fifth interview are face-to-face interviews; the remaining interviews are conducted, if possible, by telephone. The design of the CPS implies that the FSS data are collected across mixed modes; depending on the month-in-sample, the interview may be conducted as a face-to-face or telephone interview. The respondent selected for the CPS is the person who is identified as most knowledgeable concerning the labor force status of the members of the household.

Supplements to the CPS labor force questions are included in various months; for example, in March of every year, the supplement focuses on the collection of detailed data on household income, employment, and social assistance program participation. In other months, the supplements may focus on child support payments, ownership of home computers and use of the Internet, or health-related behaviors. As noted above, the FSS is currently collected as a supplement to the CPS in December of each year. Prior to the administration of the FSS, the interviewer determines the most knowledgeable member of the household concerning food that is purchased and eaten by the household and interviews that person.

Since moving to the December field date, the rotating panel design of the CPS implies that the FSS is administered to each housing unit twice, one year apart. One caution is that, as stated above, the CPS is a housing unit–based sample; individuals and families are not followed if they move from the CPS selected housing unit. For example, a sampled housing unit for which December 2002 represented the first month-in-sample will be interviewed again in December 2003 (fifth month-in-sample). Hence, changes over a one-year interval can be examined for up to 50 percent of the sampled households that have had no change in their composition.

The reinterview of part of the sampled cases allows estimation of within-household variance for the food security measures.[1] Researchers in the Economic Research Service (ERS) of USDA are currently using the panel feature of the CPS to look at food insecurity of households as they approach the beginning of a food stamp spell, a period of one or more months during which a household receives food stamps every month. Wilde and

---

[1] By applying appropriate statistical methods, one can account for the intrahousehold variations in food security in the estimated distribution of food insecurity in the population although the sampled household may have been replaced by another household at the same address.

Nord (2005) used the 2002–2003 CPS Food Security Supplement panel to estimate the effect of food stamp program participation on food security status. That paper points to some of the inherent problems and suggests some directions for future research.

The Food Security Supplement

The Food Security Supplement to the CPS includes many questions in addition to the core HFSSM. As described in Chapter 2, it contains a battery of more than 70 questions regarding various aspects of household food use and experiences during the 30 days and 12 months prior to the interview. Appendix A contains the FSS questions (note: not all the follow-up questions are numbered in the set of questions shown in Appendix A). The FSS includes five major sections:

1. Food expenditures.
2. Minimum food spending needed.
3. Food assistance program participation.
4. Food sufficiency and food security (this section includes the 18 food security and hunger questions that are used to calculate the household food security scale).
5. Ways of coping with not having enough food.

The content of the FSS varied somewhat in 1996 and 1997, and more substantial revisions in screening and format were introduced in 1998. As described in Chapter 2, households with incomes higher than 185 percent of the federal poverty line and who give no indication of problems with food access or adequacy on either of two preliminary screening questions are deemed to be food secure and are not asked subsequent questions. The two screening questions are:

1. People do different things when they are running out of money for food in order to make their food or their food money go further. In the last 12 months, since December of last year, did you ever run short of money and try to make your food or your food money go further? (This is the first question in the food assistance program participation section in the FSS and Q. 12 in Appendix A.)
2. Which of these statements best describes the food eaten in your household—enough of the kinds of food we want to eat, enough but not always the kinds of food we want to eat, sometimes not enough to eat, or often not enough to eat? (This is the first ques-

tion in the food sufficiency and food security section in the FSS and Q. 22 in Appendix A.)

As stated in Chapter 2, the HFSSM that is the basis of the food security scale has remained constant in all years. The module consists of 10 questions for households with no children and 18 questions for households with children. These questions inquire about experiences and behaviors of households having difficulty meeting their food needs. These questions focus on:

- whether the household experienced anxiety over the lack of resources to meet basic food needs,
- insufficiency in quality of food, and
- reduced food intake or the feeling of hunger.

The questions range in severity of the food security experience from the least severe, of worrying about being able to afford food, to the most severe, of skipping or cutting meals or losing weight because of lack of food. Each question asks about a specific time frame of either the past 12 months or the past 30 days. Separate scales are developed for the different reference periods, although only the 12-month scale is commonly used in household food security analyses. The questions that comprise the household food security scale are shown in Box 2-1. Most of these questions have follow-up questions that are not included as part of the 18 questions used to assess food insecurity.

Among households responding to the items shown in Box 2-1, the classification as to food security depends on whether the household includes children. The classification categories and the number of affirmed items necessary to be in a given category are shown in Box 4-1.

A primary purpose of the food security measures is to estimate the prevalence of food insecurity in the United States. The U.S. Department of Agriculture (USDA) publishes a report each year summarizing the results of the latest round of the Food Security Supplement.[2] Table 4-1 provides estimates of the percentage of households and individuals who are food secure, food insecure without hunger, and food insecure with hunger for the years 1998–2004 based on the CPS survey.

---

[2]The latest in the series is Nord, Andrews, and Carlson (2005b).

Box 4-1
Categorization of Food Security Status of Households
According to the Number of Affirmed Items on the
Food Security Scale

Households without children (based on responses to the 10 adult and household items):

Food secure = households that denied all items or affirmed 1 or 2 items

Food insecure without hunger = households that affirmed 3, 4, or 5 items

Food insecure with hunger = households that affirmed 6 or more items

Households with children (based on responses to all 18 items):

Food secure = households that denied all items or affirmed 1 or 2 items

Food insecure without hunger = households that affirmed 3 to 7 items

Food insecure with hunger = households that affirmed 8 or more items

---

RELATIONSHIP BETWEEN CONCEPTS AND QUESTIONS

The validity of the questions used can be assessed in many ways. Content validity refers to the degree to which the items currently included in a scale represent the various facets of the concept to be measured (see Bohrnstedt, 1983)—in this case, the components of food security, food insecurity, and food insecurity with hunger. This section considers the relationships among the three major categories of food insecurity and hunger (whether the household experienced uncertainty, insufficiency in quality of food, or reduced food intake or the feeling of hunger).

1. *Household experience of uncertainty and food depletion* (Questions 1 and 2, Box 2-1). The first area of inquiry concerns whether or not the respondent has experienced an "anxiety or perception that the household food budget or food supply was inadequate" (Bickel, Nord, Price, Hamilton, and Cook, 2000). The questions ask about worrying if the food would run out before they got money to buy more, and whether what they bought just

TABLE 4-1 Prevalence of Food Security, Food Insecurity Without Hunger, and Food Insecurity with Hunger, by Year (Percentage)

| Unit | Food Secure | Food Insecure | |
		Without Hunger	With Hunger
Households			
1998	88.2	8.1	3.7
1999	89.9	7.1	3.0
2000	89.5	7.3	3.1
2001	89.3	7.4	3.3
2002	88.9	7.6	3.5
2003	88.8	7.7	3.5
2004	88.1	8.0	3.9
All individuals (by food security status of household)[a]			
1998	86.5	9.8	3.7
1999	88.5	8.6	2.9
2000	87.9	9.0	3.1
2001	87.8	8.9	3.3
2002	87.5	9.1	3.4
2003	87.3	9.3	3.4
2004	86.8	9.5	3.7
Adults (by food security status of household)[a]			
1998	88.8	7.9	3.3
1999	90.5	7.0	2.5
2000	89.9	7.3	2.8
2001	89.8	7.3	3.0
2002	89.5	7.5	3.0
2003	89.2	7.7	3.1
2004	88.7	7.9	3.4

did not last and they did not have money to get more. For both these questions, the respondent was asked if that was often, sometimes, or never true in the last 12 months.

2. *Insufficiency in quality or quantity of diet* (Questions 3, 11, and 12, Box 2-1). With respect to nutritional adequacy, the HFSSM includes items concerning not being able to afford to eat "balanced meals" for the household and the children. The HFSSM also assesses whether the household relied on "a few kinds of low-cost food" for the children; a parallel item is not included for adults in the household.

3. *Reduced food intake or the feeling of hunger* (Questions 4, 5, 6, 7, 8, 9, 10, 13–18, Box 2-1). Questions on reduced food intake, skipping meals, and going a whole day without food are asked of adults and children sepa-

TABLE 4-1 Continued

Unit	Food Secure	Food Insecure	
		Without Hunger Among Children	With Hunger Among Children
Households with children			
1998	82.4	16.7	0.9
1999	85.2	14.2	0.6
2000	83.8	15.5	0.7
2001	83.9	15.6	0.6
2002	83.5	15.8	0.7
2003	83.3	16.1	0.5
2004	82.4	16.9	0.7
Children (by food security status of household)[a]			
1998	80.3	18.7	1.0
1999	83.1	16.2	0.7
2000	82.0	17.2	0.8
2001	82.4	16.9	0.6
2002	81.9	17.3	0.8
2003	81.8	17.6	0.6
2004	81.0	18.2	0.7

NOTE: Percentages calculated by the Economic Research Service using data from the August 1998, April 1999, September 2000, December 2001, December 2002, and December 2003 Current Population Survey Food Security Supplements.

[a]The food security survey measures food security status at the household level. Not all individuals residing in food-insecure households are appropriately characterized as food insecure. Similarly, not all individuals in households classified as food insecure with hunger, nor all children in households classified as food insecure with hunger among children, were subject to reductions in food intake or experienced resource-constrained hunger.

SOURCE: Nord, Andrews, and Carlson (2005b).

rately. The adult questions also ask whether the individual has lost weight. As discussed in Chapter 3, food insecurity is a concept that refers to the social and economic problem of lack of food due to economic deprivation. As such, the concept implies decision making and allocation of resources at the consumer unit, most often the family or the household. The concept implies that the measurement or operationalization of the concept is appropriately applied at the household or family level. Hunger, in contrast, is a distinct concept that is a possible consequence of food insecurity, but it is experienced at the individual level (e.g., painful sensation). Thus, the concept is distinct from food insecurity and the unit of observation is different. Although questions concerning the sensation of hunger are included in the HFSSM, the only adult for whom the information is collected is the respondent, and, with respect to children, it is collected by proxy for all children in the household as a group.

The Concept of Balanced Meal

The concept of "balanced meal" is included in the measurement of the concept of food insecurity based on earlier ethnographic work by Radimer, who describes interviewees expressing concerns about not having specific foods or food groups (as cited in Derrickson, Sakai, and Anderson, 2001). However, evidence from telephone interviews of charitable food recipients in Hawaii as well as previous qualitative work in initial cognitive testing with low-income food "gatekeepers" (who purchased and/or prepared food) shows that the interpretation of the concept of "balanced meal" is neither valid nor reliable as a measure of food insecurity (Derrickson et al., 2001).

Frequency and Duration of Food Insecurity

As stated in Chapter 3, frequency and duration are important elements for USDA to consider in the concept and operational definition of household food insecurity. USDA's food security scale measures the severity of food insecurity in surveyed households and classifies their food security status during the previous year. The frequency of food insecurity and the duration of spells of insecurity are not assessed directly in the HFSSM questions used to classify households by food security status. Although some of the response options do offer choices of "often, sometimes, or never," these response options are not sufficient measures of frequency. In addition, the FSS also includes questions about duration. The questions ask the respondent to estimate the number of days (within the past 30) the phenomenon or behavior was experienced (e.g., in the last 30 days, how many days were you hungry but didn't eat because you couldn't afford enough food?). These questions are not used in the 18-item HFSSM, although they have been used in research to estimate the percentage of the population that is food insecure on a given day in a given month.

A recent research study undertaken by researchers at the Economic Research Service examined the extent to which food insecurity and hunger are occasional, recurring, or frequent in U.S. households that experience them (Nord, Andrews, and Winicki, 2002). The study analyzes the supplementary data along with the scale and its constituent items using data from the August 1998 CPS supplement. The study found that, on the basis of reported frequency of occurrence of individual items, about two-thirds of the food insecurity and hunger conditions measured by the 12-month scale occurred in 3 or more months of the year. Furthermore, for about one-fifth of the households that experience conditions indicating food insecurity and one-quarter of those reporting hunger-related conditions, occurrences of the reported experiences and behaviors were frequent or chronic, that is, they were reported to have occurred "often" or "in almost every month." The monthly prevalence of resource-constrained hunger as currently measured by USDA is estimated to have been about 60 percent of the annual prevalence. Among households that reported food insecurity with hunger during a month, around 70 percent of the respondents experienced these conditions in 7 days or fewer; 10 to 20 percent experienced the conditions in 15 days or more. On a typical day in 1998, the prevalence of food insecurity with hunger is estimated to have been 13 to 18 percent of the annual prevalence.

The researchers note that although "the analyses add considerably to the understanding of the frequency and duration of food insecurity and hunger, they should be used with some caution. The proportions of food insecurity and hunger that are recurring and frequent or chronic almost certainly vary across subpopulations. . . . Precise estimation of frequency or chronic incidence would require collection of frequency-of-occurrence information for all items in the scale and creation of a separate scale of frequent or chronic food insecurity and hunger" (Nord et al., 2002, p. 200).

Recommendation 4-1: USDA should determine the best way to measure frequency and duration of household food insecurity. Any revised or additional measures should be appropriately tested before implementing in the Household Food Security Survey Module.

QUESTION DESIGN ISSUES

The focus of this section is on the validity and reliability of the individual questions that comprise the HFSSM. The specific issues are summarized from Dykema and Schaeffer (2005).[3] Although the focus is on improving specific items in the HFSSM, the issues outlined in the following discussion are also applicable to the full FSS.

Measurement and Item Construction

Defining Constrained Economic Resources

The concept of constrained economic resources is not consistently referenced in the current set of questions. The focus of the questions is on limited intake and availability of food due to constrained economic re-

---

[3]This information is drawn from the background paper by Dykema and Schaeffer (2005).

sources. It is therefore critical for respondents to understand that each of the questions is limited by the constrained economic resource condition. The questionnaire uses several different terms to describe these constraints, including "before we got money to buy more," "didn't have/wasn't enough money," "couldn't afford," and "running out of money." Providing reference to the condition in a consistent manner throughout the supplement, in contrast, would reduce the burden on the respondent. In addition, the concept should be introduced to the respondent prior to the questions so as to frame the full set of questions.

Specification of the Reference Person(s)

Cognitive research indicates that questions should be grouped by the topic and reference person of interest. Throughout the FSS, including the HFSSM, the reference unit shifts among the household, the children in the household, the adults in the household, and the reference person. In addition, for many of the questions, the reference person or persons is ambiguous.

Both conceptually and analytically, these shifts in the reference person present problems for the respondent as well as in the interpretation of responses. Since food security or insecurity is a household-level phenomenon, the operationalization of the concept should be reflected in household-level questions. Similarly, if the concept of insufficient intake or hunger is an individually experienced phenomenon, questions should address the individual.[4]

In addition to the conceptual and analytic issues noted above, the specification of the reference person in the Food Security Supplement poses several challenges for respondents:

- *The unit referred to by the question changes across questions.* The unit is variously the household (e.g., "Was that often true, sometimes true, or never true for you/your household?"), adults in the household (e.g., "In the last 12 months, did (you/you or other adults in your household) ever cut the size of your meals or skip meals because there wasn't enough money for food?"), the respondent (e.g., "In the last 12 months, since December of last year, did you ever eat less than you felt you should because there wasn't enough money for food?" and "In the last 12

---

[4]Although a number of questions ask specifically about the experience of the respondent, the data cannot be used to produce unbiased estimates because the respondent to the supplement is not randomly selected.

months, since December of last year, did you ever cut the size of (your child's/any of the children's) meals because there wasn't enough money for food?"), and children within the household (e.g., "In the last 12 months, did (your child/any of the children) ever skip a meal because there wasn't enough money for food?"). Moreover, the questions (as administered) are not grouped by reference unit but move among the household, the children, adults in the household, and the respondent.

- *The unit referred to changes within question.* For example, one of the questions presents the following to the respondent:

  "The food that (I/we) bought just didn't last, and (I/we) didn't have money to get more. Was that often, sometimes, or never true for you in the last 12 months?" In households with more than one adult, the statement uses the plural "we," but the target question asks about "you."

Although some respondents may infer that "you" refers to "your household," others may simply shift focus to the easier unit to report about, the respondent himself or herself.

- *The unit is an aggregate unit that implicitly requires that the respondent aggregate or summarize in order to describe the unit.* In responding to the statement "(I/We) worried whether (my/our) food would run out before (I/we) got money to buy more. Was that often true, sometimes true, or never true for (you/your household) in the last 12 months?" the respondent must summarize the experience of multiple adults. These experiences may be variable within household—one person may worry sometimes, another may never worry. It is not clear whether the respondent should report that "your household" worries if at least one adult worries or only if all adults worry. Similarly, the question assumes that there is some economic sharing among the adults in the household, so that it makes sense to say, "before we got money to buy more." It is not clear how many adults would need to "get money to buy more" and which adults they must be willing to share the food they buy with.
- *Many questions ask for proxy reports.* The respondent may not know how often all the adults in the household cut the size of meals or skip meals. If the respondent is not the person who feeds the children, she or he may not know how often the size of the children's meals was cut. Several of the questions in the HFSSM and the FSS appear to use "you" as singular to refer to the respondent themselves (e.g., "did you ever eat less than you

felt you should because there wasn't enough money for food,"
"were you ever hungry but didn't eat because you couldn't afford enough food?") This seems appropriate, because the experiences asked about seem unsuited to proxy reports; if "you" in
these questions refers to the respondent, then it is inconsistent
with the use of "you" as (probably) plural in other FSS questions. The experience of "worry" ("We worried whether our
food would run out") also seems unsuited to proxy reports, but
it is asked about as though it is a household-level concept.

Reference Periods

A large body of empirical literature exists that examines the relationship between length of the reference period and the level of measurement
error (see for example, Bound, Brown, and Mathiowetz, 2001). As noted
by Schaeffer and Presser (2003), "The choice of reference period is usually
determined by the periodicity of the target event, how memorable or pat-
terned the events are likely to be, and the analytic goals of the survey."

In principle, the 12-month reference period currently employed in the
FSS allows researchers to estimate the proportion of households that experienced food insecurity in a year, a concept that encompasses seasonal variation. While the motivation to use a 12-month reference period is clear,
evidence indicates that the reference period may be too difficult for respondents to implement accurately. For example, analysis of prevalence rates of
food insecurity and hunger suggested a seasonality effect such that rates
differed depending on whether the survey was fielded in the spring (April)
or the fall (September) (Cohen et al., 2002). Although one might expect
that episodes of severe food shortage or hunger would be salient and therefore memorable, it is possible that the occurrence of such events is salient,
but their frequency, duration, or timing are not reported accurately. This
could occur if episodes of hunger occur in stressful contexts that are not
conducive to encoding these experiential features. Furthermore, episodes of
severe food shortage or hunger (or of constrained resources) may have vague
boundaries: it may be easy to say that a time of constrained resources has
now become a time of hardship, but it may be difficult to pinpoint exactly
when the character of the event changed. Such ambiguities may make it
difficult to encode or enumerate events like times of hunger.

Two important issues regarding the reference periods included in the
questions need to be addressed: (1) the appropriateness of using (for the
most part) an annual reference period to evaluate food insecurity and (2)
the best way to specify the reference period within the wording of the individual questions. Note that the FSS as administered to the respondent includes not only the 12-month reference period but also 30-day follow-up

questions (e.g., did this ever happen in the last 30 days). Therefore, in the administration of the questionnaire, the respondent is constantly being asked to shift between a 12-month and a 30-day reference period. Such shifts are cognitively difficult for respondents and should be avoided if possible.

The 12-month reporting period used in the HFSSM questions could reduce the reliability of responses. Findings from an evaluation study conducted after the 1995 survey pretest (Hess and Singer, 1995) indicated that nearly 25 percent of the respondents "failed to understand correctly the time period referred to by the question asking about 'the last 12 months'."

Moving to a 30-day reference period may reduce the cognitive burden of recalling phenomena for a 12-month period, but, as expected, it could also reduce the percentage of households classified as food insecure. A 30-day scale was originally developed for use in the analysis of the 1995 CPS data, but it has not been used much. Researchers at ERS have revised the scale to make it more consistent with the standard 12-month U.S. food security scale commonly used in food security household analyses. A nonlinear (Rasch model–based) scaling method was used to statistically assess both the original and revised scales (Nord, 2002b). The report of this work examines the feasibility of a 30-day food security scale, and it specifies procedures for calculating the revised 30-day scale from the FSS data and classifying households as to 30-day food security status. It also compares prevalence rates of food insecurity with hunger based on the 30-day scale with those based on the 12-month scale for the years 1998–2000. Nord found that of the 3 percent of households classified as food insecure with hunger based on a 12-month scale, 74 percent (or 2 percent of the households) were similarly classified for the 30-day period prior to the survey.[5]

The impact of the length of the reference period and confusion about its boundaries on the quality of the resulting data depends partly on the actual organization of the episodes of food insecurity in respondents' lives. Both the difficulty of the reporting task and the organization of the respondents' experiences will influence which heuristics respondents use to supplement their memory in constructing answers. For example, a respondent who experiences food shortages monthly because of the timing of income may take a monthly value and multiply by 12 to produce an answer. Another example is suggested by Bickel and colleagues (2000, p. 16), "The U.S. standard food security measure reflects the household's situation over the 12 months before the interview. . . . A household that experienced food insecu-

---

[5]In addition, given the current design in which the FSS is asked only as a supplement to the December CPS, a 30-day reference period would limit analysts to examining the period between mid-November and mid-December with respect to food insecurity.

rity at some time during the past year (or other period), and therefore is considered food insecure, may in fact be food secure at the time of the interview." In such a case, the respondent's beliefs about the stability or level of change in periods of food insecurity or hunger may supplement the respondent's memory as the answer is constructed (see, for example, Ross and Conway, 1986). Other heuristics that supplement or replace memory when answers are constructed are described in Schwarz (1994) and Tourangeau, Rips, and Rasinski (2000).

The items currently included in the HFSSM vary in how the reference period is specified and where in the question the reference is located. In addition, the reference period confusion is exacerbated by the manner in which the questions are actually administered.

Response Scales

Items in the HFSSM use several formats to record the frequency of the various psychological states or behaviors. Six of the questions in the HFSSM require respondents to rate how often (i.e., often, sometimes, or never) the behavior or psychological state in question was true for them (or in some sort of aggregation across adults in the household). Three follow-up questions ask respondents to assess whether the behavior occurred "almost every month, some months but not every month, or in only 1 or 2 months." Several of the questions in the FSS use the response options "often true, sometimes true, never true." For the remaining items, respondents answer using a dichotomous yes/no format. When responses to scale items are converted into numerical scores, both "often" and "sometimes" responses are collapsed. The problems in deciding how often a description of an event is "true" are exacerbated when it is a complex event (so that part of the event described may have happened and part not) and a compound event (which is aggregated over multiple actors, who may have experienced the event with different frequencies). The use of vague quantifiers (often, sometimes) further confounds the interpretation of the response options.

Interpretation of Question Wording

One of the critical concerns in designing questions is ensuring that respondents interpret terms consistently. Cognitive testing on low-income respondents in upstate New York who were mostly white and black found that most respondents understood the terms of "hungry" or "not eating enough" as intended ("hunger" as a severe problem of decreased food quantity and "not eating enough" as less severe), however, some respondents also associated reduced quality with "not eating enough" (Alaimo, Olson, and Frongillo, 1999). As noted above, respondents may not understand the concept of "balanced meal" as applied, and therefore the question may not be understood in terms of "insufficient quality."

Conclusion

Many of the issues outlined in the preceding discussion can be classified under an overarching principle of questionnaire design, that is, the reduction of cognitive burden for the respondent. Consistent use of terminology, clustering questions so as to focus on a specific reference person or reference group (e.g., the respondent, all adults) and on a specific reference period, and developing response options that most closely map to the respondent's representation of the behavior (or attitude) are all means by which questions can be designed to reduce cognitive burden and, as a result, improve the validity and reliability of the measures. *Inevitably, questionnaire design requires balancing multiple intents and principles, and there is no perfect questionnaire design. Nevertheless, the panel concludes that the questions in the HFSSM in particular and the FSS in general can be improved by attending to these design principles as well as possible.*

Finally, the panel notes that many of the questions included in the FSS are not incorporated in the classification of households as food secure or insecure. However, they are used for research by USDA and other researchers. When developing the food security scale, Hamilton and colleagues (1997a, 1997b) tested some of these questions for inclusion in the scale but decided against using them after testing the scale. If any of the questions are not important for research purposes, they should be deleted from the FSS. However, any changes made to the supplement should be mindful of potential context effects.

In reviewing the research, the panel was impressed with the unusually comprehensive program of methodological research conducted in the mid-1990s. That series of studies provides, in many respects, a model on which to ground future research.

At the time those studies were conducted, cognitive assessment of questions was undertaken prior to the launching of the 1995 CPS supplement, but the field was not as advanced as it is now (see e.g., Willis, 2004; Presser et al., 2004). USDA should consider cognitive interviews to explore who in the household is the most appropriate person to answer the questions and what topics are appropriate for proxy responding. Following substantial cognitive testing, if a major revision of these items is undertaken, it is then appropriate to focus on improvements to the reliability of the items by simplifying them and the cognitive burden they impose.

In addition to cognitive assessment of the individual items, the use of computer-assisted interviewing (either for in-person or telephone interviews) offers a means by which to consolidate questions by topic, reference person, and reference period without overburdening the interviewer. For example, should USDA wish to maintain questions concerning both 12-month and 30-day reference periods, all the 12-month questions could be grouped together. Items that were affirmed for the 12-month reference period could be followed up, focusing exclusively on a 30-day reference period.

Recommendation 4-2: USDA should revise the wording and ordering of the questions in the Household Food Security Survey Module. Examples of possible revisions that should be considered include improvements in the consistent treatment of reference periods, reference units, and response options across questions. The revised questions should reflect modern cognitive questionnaire design principles and new data collection technology and should be tested prior to implementation.

5

Item Response Theory and
Food Insecurity

I tem response theory (IRT) models, and in particular the Rasch model,
are important elements in the U.S. Department of Agriculture (USDA)
classification of households in terms of food insecurity. This chapter
reviews IRT and related statistical models and discusses the use and applicability of IRT models in the development of such classifications. In addition, some modifications of the current IRT methodology used by USDA
are recommended that can increase the amount of information that is used
and make the methods more appropriate to the types of data that are currently collected using the Food Security Supplement (FSS) to the Current
Population Survey (CPS).

This chapter is organized as follows. The first section provides a brief
history of latent variable models, of which IRT models are a special case.
The next section discusses latent variable models in general and IRT models
in particular. It contains a description of how latent variable models are
parameterized, their interpretation, and the concept of conditional independence that drives their modeling. It also discusses how they are estimated
using data and some general issues of the identifiability of these models.
The next section considers how IRT models are used by USDA in the measurement of food insecurity and suggests how they might be used in different (improved) ways to accomplish this measurement. The last section considers a simple way to modify the existing models currently used by USDA
to take into account the polytomous nature of the data collected. A summary of conclusions reached and the recommendations that flow from them
concludes the chapter.

BRIEF HISTORY OF LATENT VARIABLE MODELS

Statistical models that incorporate latent variables (i.e., variables that are inherently unobservable) began at least as early as the observation of Spearman (1904) that scores on different educational or academic tests were usually positively correlated; that is, examinees performing well on one academic test often performed well on other tests. This phenomenon was observed in many circumstances, and Spearman concluded that it could be explained by a simple statistical model in which each examinee was postulated as having an underlying unidimensional, but not directly observed, "academic ability" or "general intelligence" that varied from person to person. He assumed that this ability was positively related to a person's performance on each of the different tests. The higher a person's ability, the higher he or she tended to score on *any* test of some aspect of academic or intellectual performance.

Spearman's simple model was elaborated and led to the development of *factor analysis* as a statistical methodology, as well as to various theories of intelligence, as a topic within psychology. Early references to factor analysis are Spearman (1904), Thurstone (1931), and Kelley (1935).

Closely related to factor analysis was *true score theory,* in which a single educational or "mental" test was the object of study rather than several tests. In this framework, *observed test scores* were considered the result of a *latent true score* plus *measurement error*. This was a powerful theory that allowed the development of quantitative measures of reliability and validity that have become routine measures of the efficacy of any test (Spearman, 1907; Kelly, 1923).

Starting in the 1940s, *latent structure* or *latent class* models were developed and applied to sets of individual test or survey questions to produce scales for both the questions and the respondents (Stouffer et al., 1950). These were further developed in Anderson (1954) and Lazarsfeld and Henry (1968). At roughly the same time, *item response theory,* of which the Rasch model is an example, was developed for educational and psychological tests (Lawley, 1943; Tucker, 1946; Lord, 1952; Rasch, 1960; Birnbaum, 1968; Lord, 1980). The word "item" in item response theory is a term used by test developers and psychometricians to refer to the questions on tests and the rules for scoring them.

Bartholomew (1987) gives a unified discussion of the three related types of latent variable models—factor analysis, latent class analysis, and item response theory. This general class of statistical models is discussed more extensively next.

STRUCTURE OF LATENT VARIABLE MODELS

Factor analysis, latent class analysis, and item response theory when viewed as statistical models all share a basic mathematical structure. Examples from the measurement of food insecurity, when appropriate, are used to make the ideas concrete. These three types of statistical models all involve several observed variables or measurements and one (or more) latent, unobserved variable. These models have contact with data because they may be used to describe the distribution of the observed variables over a population of respondents. In addition, they allow users to draw inferences about the unobserved latent variable (e.g., food insecurity) based on the observed data (e.g., the FSS questions).

In general, the observed data consist of a set of p variables that are observed for each respondent in the study. These are called the *manifest variables*. Denote them by $X_1, X_2, \ldots X_p$. In factor analysis, the X's are the observed test scores from p tests for each person in the study. In latent class analysis, the X's are the observed categorical responses of each respondent to p questions on a survey instrument. In IRT, the X's are the dichotomous/binary or ordered polytomous responses of respondents to p questions/items on a test or survey instrument. In factor analysis, the manifest variables are continuous variables. In latent class analysis, the manifest variables may be dichotomous or polytomous *nominal* variables whose values are unordered categories. In IRT, the manifest variables are typically categorical and *ordered* and may be dichotomous/binary (e.g., "wrong/right" or "affirmed/not affirmed") or polytomous (e.g., "never, sometimes, often"), as opposed to the continuous manifest variables of factor analysis or the unordered nominal manifest variables of latent class analysis.

In addition to the manifest variables, all latent variable models also assume the existence of a *latent variable*, the value of which varies from respondent to respondent but that is not directly observable for any respondent. The value of the latent variable affects the distribution of each manifest variable for each respondent—for example, the probability of endorsing each food insecurity question. This chapter uses the symbol ϕ to denote the latent variable to remind us that the main application of interest here is the measurement of *food insecurity*.

The three different types of latent variable models make different assumptions about the nature of the latent variable and how they are connected to the manifest variables. In factor analysis, each latent factor is continuous and univariate and the mean or expected value of each manifest variable is a linear combination of the latent factors. The weights on these linear combinations indicate the influence of each underlying latent factor on each test score.

In latent class analysis, the latent variable is a discrete latent class to which each respondent is assumed to belong. Thus, in latent class analysis the latent variable is categorical rather than continuous, and it may or may not have an implied order among its values. Latent class models do not necessarily assume any particular form for the connection between the manifest variables and the latent class variable. (This is a source of the problem of identifiability discussed in a later section.)

For IRT models, the latent variable is continuous and univariate (or multivariate). In educational applications, the latent variable indicates the underlying latent proficiency of each test taker that in turn influences the likelihood of correct responses to the test. In the application to food insecurity measurement, the latent variable represents the degree of food insecurity experienced by a given household that in turn influences the likelihood of endorsing or affirming responses to questions about lack of food due to economic constraints. There is a close connection between IRT models and latent class models. If the latent variable in an IRT model is assumed to have a *discrete distribution* concentrated on a few points, it becomes a latent class model with ordered latent classes.

Deciding whether a latent variable is more appropriately thought of as discrete or continuous cannot really be based on data, and in fact it is often impossible to assess any difference between the fit of the two types of models (Lindsey, Clogg, and Grego, 1991). More usually, this decision is based on other considerations. For example, in the case of food insecurity, it seems plausible that ϕ varies in a continuous way across households rather than only having a few possible values that it can take on.

It is evident that, of these different types of latent variable models, IRT models are particularly appropriate for modeling the measurement of food insecurity using survey data of the type collected in the CPS. The manifest variables or indicators of food insecurity in the FSS are all either binary or polytomous and ordered. In addition, food insecurity may be viewed as an underlying continuous, unidimensional, but not directly observable quantity that varies from household to household. Higher values of latent food insecurity are indicated by higher probabilities of endorsing or affirming survey items that indicate higher degrees of not being able to obtain sufficient food due to a lack of economic resources.

Returning to the structure of latent variable models, they all involve the notion of *conditional statistical independence*, so the panel first reviews this important idea.

Conditional Statistical Independence

Statistical Independence

A familiar example of statistical independence is the result of two tosses of a coin. Neither coin toss can influence the outcome of the other so they yield statistically independent results. More generally, if two variables are statistically independent, then neither one affects probabilities that involve the other variable. This is a very strong notion that there is "no relationship" between the two variables. This idea is formalized using *conditional probability*, and, to define it, some notation is now introduced that applies to the rest of this chapter.

The conditional probability that one variable, X_2, has the value x_2, given that (or conditional on) the fact that another variable, X_1, has the value x_1 is commonly denoted by

$$P\{X_2 = x_2 \mid X_1 = x_1\} . \tag{1}$$

In the example of two tosses of a coin, X_1 could denote the outcome of the first toss and X_2 the outcome of the second toss. In this example, x_1 and x_2 are the values "heads" and "tails."

In terms of conditional probability, the *statistical independence* of X_1 and X_2 is expressed by

$$P\{X_2 = x_2 \mid X_1 = x_1\} = P\{X_2 = x_2\} , \text{ for any } x_1 \text{ and } x_2. \tag{2}$$

The probability on the right side of equation (2) is just the ordinary, *marginal*, or *unconditional* probability that $X_2 = x_2$. The equality of the two probabilities in equation (2) means that the probability distribution of X_2 is *unaffected* by the value of X_1. In other words, the conditional probability is *constant* as a function of x_1.

It is well known (for example, see Parzen, 1960) that the constant conditional probability rule in equation (2) is equivalent to the following "product rule" for joint probabilities of independent variables

$$P\{X_2 = x_2 \text{ and } X_1 = x_1\} = P\{X_2 = x_2\} \, P\{X_1 = x_1\}. \tag{3}$$

The product rule means that the *joint probability* that $X_1 = x_1$ *and* that $X_2 = x_2$, the left side of equation (3), is found by multiplying together the two marginal probabilities for each variable separately, the right side of equation (3). Both the constant conditional probability rule in equation (2) and the product rule in equation (3) are important for understanding the structure of latent variable models.

By itself, statistical independence is too strong a condition to apply directly to most survey data. For example, for questions on the FSS in the CPS, the assumption of statistical independence asserts that the probability that an individual endorses or affirms any survey item is independent of whether or not they endorse any other item. On its face, this assumption seems too strong, since it would be expected that the endorsement of one food insecurity question would increase the probability of endorsing other food insecurity questions. However, a modified form of independence, *conditional statistical independence*, is a more useful idea and is described next.

Conditional Statistical Independence

Two variables, X_1 and X_2, are statistically independent *conditionally given a third variable*, Z, if their probabilities satisfy a *conditional* version of the product rule in equation (3), i.e.,

$$P\{X_2 = x_2 \text{ and } X_1 = x_1 \mid Z = z\} = P\{X_2 = x_2 \mid Z = z\} \, P\{X_1 = x_1 \mid Z = z\} . \quad (4)$$

Equation (4) says that when Z is fixed at (or conditioned to be) a specific value, z, then X_1 and X_2 are statistically independent, using the product rule. It is possible for variables to *be* conditionally statistically independent given a third variable but *not to be* statistically independent themselves. In this circumstance, it is sometimes said that Z "explains" any association or dependence between X_1 and X_2, because, once the value of Z has been conditioned on or fixed, there is no more association left to explain. A coin tossing example of conditional independence arises if there are two unfair coins. For example, suppose that coin A is biased towards heads and produces heads with probability 2/3, while coin B is biased towards tails and produces heads with probability 1/3. Now the procedure is to pick one of the two coins out of a box at random and then toss it twice. If which coin is being tossed is known, then there is statistical independence between X_1 and X_2 as before. In this case Z is the coin being tossed, A or B. Conditioning on or knowing which coin was selected makes the results of the two tosses be independent. But if the coin being tossed is unknown, then X_1 and X_2 are statistically dependent. If, for example, the coin is pulled out of the box at random and tossed and X_1 is heads, then it is more likely than not that the coin is A, and therefore X_2 is more likely than not to be a heads as well.

Measurement Models

One way to understand the role of conditional independence in latent variable models is in terms of *measurement models*. In this usage, the latent variable, ϕ, is regarded as the quantity being "measured." The value of ϕ for a given respondent is regarded as fixed, and the values of the various observed measurements or manifest variables, the **X**'s, for that respondent are regarded as statistically independent *indicators* of different aspects of ϕ. The connection between ϕ and the **X**'s is assumed *probabilistic* so that two respondents with the same value of ϕ may still have different values for the observed values of the **X**'s. Any correlation among the manifest variables that is observed across the population of respondents is regarded as due to the fact that the manifest variables all measure the same underlying quantity ϕ that varies across respondents.

From this perspective, the conditional independence of the **X**'s given a respondent's value of ϕ is the natural way to define "independent" indicators of ϕ. The relationship between the manifest variables and the latent variable is then expressed by the equation of conditional independence of the **X**'s given ϕ, i.e.,

$$P\{X_1 = x_1, \ldots, X_p = x_p \mid \phi\} =$$
$$P\{X_1 = x_1 \mid \phi\} P\{X_2 = x_2 \mid \phi\} \ldots P\{X_p = x_p \mid \phi\} . \tag{5}$$

Equation (5) is the basic defining assumption of all latent variable models. It says that the joint distribution of the manifest variables simplifies to independence once one conditions on the latent variable, ϕ. In IRT, this is called the assumption of *local independence*. Equation (5) is couched under the assumption that the manifest variables are discrete rather than continuous because the application to food insecurity involves such data. In the language of educational and psychological measurement, the form assumed for the conditional probabilities, $P\{X_i = x_i \mid \phi\}$, is often referred to as the *measurement model*. These models are discussed more in a later subsection.

When applied to the measurement of food insecurity, the conditional independence specified in equation (5) implies that, for a household with a given level of food insecurity, ϕ, the probability that they affirm one food insecurity survey question is conditionally independent of whether or not they affirm any other food insecurity question. Thus, the responses to the various food insecurity questions for a single household are regarded as statistically independent even though across all the households in the study there are obvious correlations between the responses to the different questions on the FSS. As described above, the degree of food insecurity, ϕ, as it varies across households, *explains* these observed correlations.

While equation (5) appears to be a strong assumption, this is a bit misleading. For example, equation (5) cannot be *directly* tested with the data because that would require knowledge of the value of ϕ for each respondent and, by definition, ϕ is *unobserved* for every respondent. The issue of testing IRT models is discussed further in a later section.

Measurement Models and Item Response Functions

The conditional distributions for the individual manifest variables, $P\{X_i = x_i \mid \phi\}$, specify the *measurement model* that connects each manifest variable, X_i, to the latent variable ϕ. As ϕ varies over the respondents, so does the probability distribution of X_i. In IRT, $P\{X_i = x_i \mid \phi\}$ is called the *item response function*, and in latent class models it was originally called the *trace line*. In factor analysis, $P\{X_i = x_i \mid \phi\}$ is specified by a linear regression equation that connects the latent factors to the conditional expected values of the manifest variables given ϕ.

The measurement model that defines the form of $P\{X_i = x_i \mid \phi\}$ includes parameters that may vary with the X_i's and allow the measurement model to express different aspects of the conditional distribution, $P\{X_i = x_i \mid \phi\}$. In IRT, these are called the *item parameters* because each X_i denotes the responses to one item. Two types of item parameters that arise repeatedly in IRT are:

a. Those that indicate the location of the conditional distribution, $P\{X_i = x_i \mid \phi\}$, along the ϕ-scale, called item "difficulty" parameters because the larger they are the less likely the respondent is to give response x_i.

b. Those that indicate the strength of the connection between X_i and ϕ—the item "discrimination" parameters. The larger these are, the stronger and the more deterministic the connection is between ϕ and X_i.

The Rasch Model

To make the idea of a measurement model and item parameters more concrete, consider the Rasch model used by USDA. Suppose X is the manifest variable that codes the response to a given food insecurity question as 1 = affirm and 0 = not affirm. For the Rasch model, the *item response function* is determined by the conditional probability of affirming the binary food insecurity question given ϕ, $P\{X = 1 \mid \phi\}$, and is given by the formula

$$P\{X = 1 \mid \phi\} = \exp\{\phi - b\} \,/[1 + \exp\{\phi - b\}]. \tag{6}$$

In equation (6), b is the item parameter. As b increases, the probability of affirming the question decreases for a household with a specific value of ϕ, that is, for larger values of b the question is "harder" and less likely to be affirmed by respondents. Thus, b is an example of a "location or difficulty" parameter for the item response function. As ϕ increases while b remains fixed, the probability increases. Thus, the Rasch model exhibits a monotonic increasing relationship between the latent variable and the probability of affirming the item. When ϕ is larger than b, then the probability that the household affirms the item is greater than 1/2—the respondent is more likely than not to affirm the item. When ϕ is smaller than b the reverse holds.

The Rasch model is very simple and does not have a separate parameter for the strength of the connection between ϕ and **X**. An example of a model with this additional type of parameter is the 2-parameter logistic (2PL) model whose corresponding item response function for dichotomous items is given by

$$P\{X = 1 \mid \phi\} = \exp\{a(\phi - b)\} / [1 + \exp\{a(\phi - b)\}]. \tag{7}$$

In equation (7), the "discrimination" parameter, a, must be greater than or equal to zero and indicates the strength of the connection between ϕ and **X**. When a is near zero, the connection is weak (i.e., for $a = 0$ there is no connection between ϕ and the probability (1/2) that the respondent endorses the item). When a is positive and large, the connection is strong and the item is said to be highly discriminating. For the Rasch model, the discrimination parameters are all assumed to be the same and correspond to setting $a = 1$. For the 2PL model, when ϕ exceeds b the respondent is more likely than not to affirm the item.

Threshold Models for Item Response Functions

Both the Rasch model and the 2PL model are examples of *threshold models* that are used in other applications in which observations are made with some degree of measurement error. An example in which such models are often used is the field of signal detection, in which an observer is trying to identify a signal in the midst of a noisy background (Peterson, Birdsall, and Fox, 1954; Birdsall, 1955). Threshold models provide a simple explanation for the form assumed for the item response functions in equations (6) and (7) that is described next.

Continuing the example of a dichotomous/binary item response of "affirming or not affirming" a given item on the FSS, a threshold, b, is assumed such that if the value of ϕ for a household exceeds b then the respondent will affirm the item, but if ϕ is below b then a nonaffirming response is given.

This measurement model has a deterministic connection between ϕ and the response. To introduce measurement error to make it more realistic, instead of ϕ determining the responses directly depending on where ϕ is relative to b, ϕ is first perturbed by a stochastic measurement error, V/a, and then the same rules for determining responses to the item are applied to

$\phi + V/a$ rather than to ϕ. The item parameter, a, determines how much the measurement error changes ϕ. The parameter a is the previously mentioned item discrimination parameter from equation (7) in a different guise, just as the threshold b is the item difficulty parameter mentioned earlier. A large a-value results in little measurement error, and a small a-value indicates a large amount of measurement error. The stochastic quantity V is taken to be independent of ϕ and to have a symmetric distribution with mean 0. Hence, the value of $\phi + V/a$ fluctuates around ϕ in a random way.

The probability of affirming the item, the item response function, is given by:

$$P\{X = 1 \mid \phi\} = P\{\phi + V/a > b \mid \phi\} = P\{V > a(b - \phi) \mid \phi\} = F_V(a(\phi - b)), \qquad (8)$$

where, in equation (8), $F_V(t)$ denotes the cumulative distribution function (cdf) of V. For the Rasch and 2PL models, $F_V(t)$ is assumed to be the logistic cdf, while for the Normal Ogive model it is the Gaussian cdf.

Finally, the conditional independence assumption in equation (5) corresponds to the assumption that the measurement errors for different items, V_i, are statistically independent. Threshold models provide a convenient way to fit many types of latent variable models into a common framework.

Only two of the items on the Household Food Security Survey Module (HFSSM) are actually dichotomous or binary, requesting a yes or no response. The other questions are either trichotomous or are two-part questions that, when considered together, have four possible ordered responses. The current use by USDA is to reduce the nondichotomous item response to binary responses by collapsing the response options to two possibilities that are regarded as either *affirming* or *not affirming* the question. A later section briefly considers more general item response functions that are directly applicable to the case of polytomous ordered responses to the food insecurity questions.

The Latent Distribution

In order to be able to specify the joint distribution of the manifest variables, $X_1, X_2, \ldots X_p$, it is necessary to integrate out or marginalize over the *latent distribution*, $f(\phi)$. Depending on the continuous or discrete nature of ϕ, $f(\phi)$ is assumed to be either a probability density or a discrete probability function. The latent distribution reflects the heterogeneity of ϕ across the population of respondents at a relevant point in time. A common assumption for IRT models is that $f(\phi)$ is the Gaussian distribution with mean 0 and variance 1. However, the latent distribution may have parameters that describe both the location and the degree of variation in ϕ over a particular population of respondents as well. Johnson (2005) suggested a left-truncated

Gaussian distribution for the case of food insecurity because the full set of food insecurity questions is asked only for those households that are likely to have large values of ϕ.

The latent distribution, $f(\phi)$, along with the item response functions, $P\{X_i = x_i \mid \phi\}$, may be combined using equation (5) to specify the joint distribution of the manifest variables, i.e.,

$$P\{X_1 = x_1,\ldots,X_p = x_p\} = \int P\{X_1 = x_1|\phi\}\ldots P\{X_p = x_p|\phi\}f(\phi)d\phi, \tag{9}$$

for continuous latent variables; a sum replaces the integral in equation (9) for discrete latent variables.

The parameters of the joint distribution of the **X**'s in equation (9) include both the item parameters from the item response functions and possibly other parameters from the latent distribution. It is this joint distribution for the manifest variables that allows these parameters to be estimated and for the latent variable model to be tested against data.

Multiple Groups of Respondents

It often happens that important subgroups of respondents need to be studied separately. For example, households with children are asked questions that are not appropriate for households without children. It is possible for the latent distribution to vary with the subgroup. When there are large differences in these latent distributions, it may be important to include them in the model for the manifest variables.

To denote this situation, let **G** be a variable that distinguishes between different subgroups of respondents. For example **G** = 0 could indicate a household *without* children, while **G** = 1 indicates a household *with* children. In this setting, equation (9) can be expanded to

$$P\{X_1 = x_1,\ldots,X_p = x_p|G = g\} =$$
$$\int P\{X_1 = x_1|\phi\}\ldots P\{X_p = x_p|\phi\}f(\phi|G = g)d\phi, \tag{10}$$

where **G** = *g* denotes one of the subgroups of interest. In many IRT applications, the latent distributions, $f(\phi \mid G = g)$, are assumed to be Gaussian with means and variances that vary with *g*.

Differential Item Functioning

In order for equation (10) to be a correct formula, ϕ has to "explain" (in the sense of conditional independence mentioned earlier) any dependence between subgroup membership and responses to the questions. In other words, given ϕ, both X_i and G must be conditionally independent for any item.

It is possible for this type of conditional independence to fail. When X_i and G are *not* conditionally independent given ϕ, the item indicated by X_i is said to exhibit *differential item functioning* (DIF), that is, to function differently for some or all of the subgroups indicated by G. For example, some of the FSS questions regarding adults could be easier to affirm for adults in households with children than for those without children. This might occur if the available household resources were used to feed the children *first* while the adults went without food for some period. In this hypothetical example, the item would function differently in terms of ϕ for households with and without children. Differential item functioning is discussed extensively in Holland and Wainer (1993). If an item is found to exhibit DIF in two groups, it may be appropriate to allow different item parameters for it that depend on which subgroup the respondent is in rather than the usual *no-DIF* assumption that they are the same for all subgroups. USDA should evaluate the amount and consequences of DIF in the FSS.

The Latent Posterior Distribution

It is evident that what can be deduced about the latent variable for a respondent depends on the values of the manifest variables for that respondent and the features assumed for the latent variable model. In particular, the *latent posterior distribution*, denoted here by $f(\phi \mid X_1 = x_1, \ldots, X_p = x_p)$, of ϕ given the values of the manifest variables, $X_1 = x_1, \ldots, X_p = x_p$, summarizes everything that is known about the value of ϕ for any respondent with a given pattern of values of the manifest variables, $X_1 = x_1, \ldots, X_p = x_p$. When the household's membership in a subgroup, indicated by $G = g$, is also considered to be important, then what is known about f can be summarized by the latent posterior distribution, $f(\phi \mid X_1 = x_1, \ldots, X_p = x_p, G = g)$, where the conditioning is now on both the pattern of values of the manifest variables and the group membership indicated by G.

It is well known that the latent posterior distribution is given by the expression

$$f(\phi \mid X_1 = x_1, \ldots, X_p = x_p) = [P\{X_1 = x_1 \mid \phi\} \ldots P\{X_p = x_p \mid \phi\} f(\phi)] / P\{X_1 = x_1, \ldots, X_p = x_p\} . \tag{11}$$

The numerator of equation (11) is the product of the item response functions and the latent distribution, which can be estimated once the item parameters are estimated. The denominator of equation (11) is the joint distribution of the manifest variables given in equation (9).

When group membership, indicated by $\mathbf{G} = g$, is included along with the pattern of manifest variables, and it is assumed that there is no DIF as described above, then the latent posterior distribution is given by the expression:

$$f(\phi \mid \mathbf{X}_1 = x_1, \ldots, \mathbf{X}_p = x_p, \mathbf{G} = g) = [P\{\mathbf{X}_1 = x_1 \mid \phi\} \ldots P\{\mathbf{X}_p = x_p \mid \phi\} \, f(\phi \mid \mathbf{G} = g)] / P\{\mathbf{X}_1 = x_1, \ldots, \mathbf{X}_p = x_p \mid \mathbf{G} = g\}. \tag{12}$$

The numerator and denominator of equation (12) are similar to those of equation (11) but now include conditioning on the value of $\mathbf{G}$ as well.

The equations for the latent posterior distributions given in equations (11) and (12) show how the various features of the latent variable model combine. These features connect what is observed about the respondents in terms of the manifest variables to the unobservable latent variable ϕ. When the latent posterior distribution is concentrated around a small interval of ϕ-values, the manifest variables restrict what the value of ϕ can be. The stronger the dependence of the item response functions on the value of ϕ and the more of them there are, the greater the restriction on the value of ϕ and the more that is known about it.

Estimating IRT Models

The parameters of IRT models may be estimated several different ways using the observed data from the manifest variables. These include the method of moments (Lord and Novick, 1968), conditional maximum likelihood (Andersen, 1980), joint (or unconditional) maximum likelihood (Lord and Novick, 1968), maximum marginal likelihood (Bock and Lieberman, 1979), and fully Bayesian methods (Patz and Junker, 1999; Bradlow et al., 1999; Johnson, 2004). Each of these methods has both strengths and limitations.

The method of moments estimates are simple and direct but appear to be feasible only for Normal Ogive models that are assumed to have a Gaussian latent distribution. Conditional maximum likelihood estimation eliminates the need for assumptions about the latent distribution but is satisfactory only for the Rasch model and closely related models in which the strength of the connection between the latent variable and the manifest variable is assumed the same for all of the manifest variables. Joint maximum likelihood estimation avoids assumptions about the latent distribution by treating both the item parameters and the value of ϕ for each respondent as parameters to be estimated. However, this approach is known to produce biased estimates of item parameters when the number of items is small (Haberman, 1977; Andersen, 1980; Opsomer et al., 2002). Various methods have been proposed for correcting this bias. Maximum marginal likelihood methods are widely used but require assumptions about the latent distribution. When these assumptions hold, they produce unbiased estimates of the item parameters. In addition, these methods allow for multiple groups that have different latent distributions as in equation (10). Fully Bayesian methods also require assumptions about the latent distribution and allow for multiple groups, but they are computationally more complicated than other approaches; however, they are becoming more widely used because of their flexibility and utility.

Over the years, USDA has used several of these methods to estimate item parameters, including joint, conditional, and maximum marginal likelihood methods assuming a single group of respondents. The different methods have yielded similar estimates for the item parameters when the Rasch model is assumed. However, if USDA moves to a more general model, the method used to estimate the item parameters may make a larger difference in the results.

Latent Variable Models and Problems of Model Identifiability

Because latent variables are not directly observed, there is no direct check on whether the form that is assumed for the measurement model is correct. That is, one cannot directly check that the form that is assumed for either the measurement model or the latent distribution is correct. This has the important consequence that if the form of the measurement model, $P\{X_i = x_i \mid \phi\}$, is completely unrestricted, then it is impossible to infer anything about the latent variable. In fact, the Suppes-Zanotti theorem shows that many such unrestricted latent variable models will describe the data perfectly (Holland and Rosenbaum, 1986). This result asserts that every joint distribution of p manifest variables, no matter how complicated, can be represented as conditionally independent given a unidimensional latent variable. Moreover, this representation can be achieved in many different ways. Hence, without further assumptions on the nature of the latent variable or on the conditional distributions, $P\{X_j = x_j \mid \phi\}$, or both, the assumption of a latent variable model is vacuous and has no empirical consequences.

This somewhat subtle point is not always appreciated. Latent variable models are useful in practice because of assumptions that are made *in addition to* the conditional independence assumption in equation (5). For example, in the case of food insecurity it is natural to assume that households with higher (or lower) values for ϕ will have higher (or lower) probabilities of endorsing each of the food insecurity questions. This monotonicity in the connection between ϕ and the probabilities of endorsing any of the food insecurity questions is a natural assumption, and it is critical to the use of such models for food insecurity measurement. In the use of IRT models by USDA, the item response functions, $P\{X_j = x_j \mid \phi\}$, are specified by the Rasch model, which implies a monotonic relationship between ϕ and the endorsement probabilities.

For all latent variable models, the form assumed for $P\{X_i = x_i \mid \phi\}$ determines what can be said about the latent variable. Some common examples of the types of restrictions assumed on the form of the measurement model are as follows. In factor analysis, it is assumed that there is a *linear* regression of X_i on ϕ. In IRT, the probability of a correct response to a dichotomous item is assumed to be a *monotone increasing* function of ϕ, or a very specific type of monotone function, for example, the Rasch model, the 2PL model, or the Normal Ogive model. In latent class analysis, there is often no assumption made about the item response functions or trace, $P\{X_i = x_i \mid \phi\}$. Instead, a restriction is placed on the number of values that ϕ can have, often just two. It is known that unless some restriction is placed on the number of latent classes, a variety of problems of estimation arise. These are due to the large number of parameters being estimated relative to the amount of data that are available (Bartholomew, 1987). In IRT modeling, it is also known that estimation problems arise when the assumed form of the item response functions has certain types of parameters beyond the basic location and discrimination parameters indicated in an earlier section, for example, guessing parameters (Lord, 1980). Both of these types of estimation problems are ultimately due to the lack of identifiability of latent variable models when the conditional distributions, $P\{X_i = x_i \mid \phi\}$, are not restricted sufficiently.

Thus, despite the appeal of latent variable models as ways of organizing independent indicators of an underlying latent variable, there is a clear sense that one gets out of a latent variable model no more than what one puts into it.

The issue of testing the fit of IRT and other latent variable models is complex because it is equation (5) *plus* whatever other assumptions are made about the latent variable model that must be tested, rather than each of them separately. This is discussed in more detail in Holland and Rosenbaum (1986); Stout et al. (1996) also provide methods for testing some aspects of IRT models.

Even when the restrictions placed on the latent variable model are sufficient to resolve the identifiability problem just described, there is still a slight problem of identifiability that is well known to users of *continuous* latent variables. The location and scale parameters of the latent distribution are confounded with the difficulty and discrimination parameters of the measurement model. The indeterminacy of IRT scales is avoided in a variety of ways. A common way is to assume that the mean of the latent distribution is 0 and the variance is 1, thus fixing the scale to be in terms of standard deviation units. When there are multiple groups, this restriction is placed on only one of them, the standard group, and the other groups have means and variances that are relative to those of the standard group.

USE OF ESTIMATED IRT MODELS FOR
MEASURING FOOD INSECURITY

This section examines how USDA currently uses an IRT model to measure food insecurity and its prevalence in the United States, describing its classification system for households without children, and then interpreting this in terms of the material developed earlier. For households with children similar comments apply, so this case is not considered further here.

The responses to the 10 "adult" food insecurity questions are dichotomized and then the number of affirming responses to these 10 questions is used to classify the households into food insecurity levels. Households affirming two or fewer questions are classified as *food secure*; those with three to five affirmations are classified as *food insecure without hunger*; and those with six or more affirmations are classified as *food insecure with hunger*.

Basing the levels of food insecurity on the number of affirming responses to the dichotomized questions greatly simplifies the task of classifying households. For example, using the 10 dichotomized questions for households without children, the number of possible affirmations range from 0 to 10. On the one hand, using these same 10 questions, and, accounting for the fact that two pairs of them allow only three possible responses for each pair rather than four, there are $2^6 3^2 = 576$ different possible patterns of affirming and nonaffirming responses. If the questions were not dichotomized or if missing data were also taken into account, then the number of possible response patterns across the items could be much greater.

On the other hand, this simplification has also led to criticism of the food insecurity categories. Bavier (2004) noted that households that did not affirm the one question that specifically asks ". . . were you ever hungry . . ." would be, on the basis of affirming six or more other questions, classified by USDA as *food insecure with hunger*. Furthermore, households that did affirm the "hunger" questions but affirmed only four additional questions would not be classified as food insecure with hunger. It can also be argued that the location of the "were you ever hungry" question on the Rasch scale is a more reasonable cut point for the "food insecure with hunger" category. To the right of that point, the Rasch model predicts that the respondents are more likely than not to affirm the hunger question, but to the left they are less likely than not to affirm it. The "were you ever hungry" question is the seventh least likely to be affirmed of the 10 household adult questions. This suggests that the criterion for the category should be seven or more affirmations rather than only six. Of course, the criticism of Bavier mentioned above would still apply to this cut point as well, although it would occur less often.

To estimate the prevalence of the food insecurity categories, the food insecurity classification of a sampled household is treated as a *household characteristic*, just like other characteristics, such as the number of adults in the household. The population estimates of the percentage of households in the target population that are in the three categories are computed from the characteristics of the sampled households in the standard way (appropriate weighted averages).

Thus, the estimates of the prevalence of food insecurity for households without children are simply estimates of the percentage of households in the target population that would affirm two or fewer, from three to five, or six or more of the food insecurity questions if they were asked. There is no need to use an IRT model to justify these estimates.

The role of the IRT model is to assist in other decisions. One of these is the location of the cut points that define the food insecurity categories. There are several ways to use IRT models to do this. Ohls, Prakash, Radbill, and Schirm (1999) give a brief description of the procedure to define the original set of three cut points that were later reduced to two. That discussion suggests that the focus was on deciding how many questions should be affirmed in order to classify households into the categories. This did not involve the IRT model except as justification for reducing the aforementioned 576 possible response patterns to a much smaller number of affirmations that ignores which questions were affirmed.

The alternative is to use the IRT model to set cut points along the latent ϕ-scale first and then to use these cut points to classify households in terms of their location along the latent food insecurity ϕ-scale.

Setting cut points along the ϕ-scale: The IRT model can relate the probability of affirming each question to locations along the ϕ-scale. For example, as indicated earlier, the locations of the estimated b-parameters for the items show where respondents are as likely as not to affirm each question. Moreover, the estimated Rasch item response function in equation (6) can be used to indicate where, along the ϕ-scale, respondents have a probability of, say, 80 percent, of affirming a given question. This response probability information can be used, along with expert judgment, to come to agreement as to where the cut points should be placed. Johnson (2005) discusses several approaches to setting cut points along a latent scale. The National Assessment of Educational Progress (NAEP) has a similar task in establishing its achievement levels (basic, proficient, and advanced) along its latent scales for various subjects and grades. These are also cut points along a latent continuum. In NAEP, these cut points were formed using a complex process that incorporated expert judgments and the results of scaling a large number of test questions onto a common unidimensional scale. It appears that substantially more information was available about examinee performance to make the judgments needed to locate the achievement levels for NAEP than was available for deciding on the cut points for the food insecurity levels (Ohls et al., 1999). The methods used for the NAEP achievement levels may usefully inform a similar process for food insecurity measurement. For example, their methods may suggest ways for USDA to incorporate health and other relevant data more closely into the assessment of the cut points on the ϕ-scale using special samples of households who answer the FSS questions as well. Johnson (2005) alludes to using extra data sources in his discussion of validity studies.

There is a somewhat nonintuitive aspect of deciding the location of cut points along a continuous scale that reflects the measurement error in IRT models. Suppose a cut point is established on the ϕ-scale, say at ϕ^*. Judges who examine the manifest data of households whose ϕ-values are well below or well above ϕ^* will tend to agree on which side of the cut point to locate the households. However, for a household near the cut point, these same judges will tend to disagree on the household's classification. Thus, somewhat against intuition, *disagreement* among judges about household classifications based on manifest data is a potential indicator of where to locate cut points along a latent scale.

Classifying households based on the manifest data: Once a latent variable model is estimated and the latent posterior distributions are available, these can be used to determine cut points along the latent scale. The probability that a household with given values of the manifest variables, $X_1 = x_1, \ldots, X_p = x_p$, has a ϕ-value above a cut point, say ϕ^*, is given by the integral

$$P\{\phi > \phi^* \mid X_1 = x_1, \ldots, X_p = x_p\} = \int_{\phi^*}^{\infty} f(t \mid X_1 = x1, \ldots, X_P = x_p)\, dt. \tag{13}$$

If membership in a subgroup denoted by $G = g$ is also considered when classifying a household, as it is for households with and without children, then equation (13) is replaced by

$$P\{\phi > \phi^* \mid X_1 = x_1, \ldots, X_p = x_p, G = g\} = \int_{\phi^*}^{\infty} f(t \mid X_1 = x_1, \ldots, X_p = x_p, G = g)\, dt. \tag{14}$$

If the probability in equation (13) or (14) exceeds 50 percent then it is more likely than not that the household does fall above f* on the f-scale, and it often makes sense to so classify the household. More complicated rules can be devised that take account of possibly different costs for errors of misclassification.

The approach outlined above is a basic way to form a classification system using an IRT model. If the Rasch IRT model is assumed, then it may be shown that the latent posterior distribution in equation (11) depends only on the *number* of affirming responses of the household rather than on which *questions* are affirmed. This simplifies the classification rule to make it more like the one used by USDA, but it requires that the Rasch model accurately represents the distribution of the observed responses to the dichotomized HFSSM questions. Johnson (2004) indicates that the 2PL model provides a better fit to the HFSSM data that he examined.

Prevalence Rates on Latent and Manifest Scales

When a cut point, ϕ^*, has been established along the ϕ-scale, the prevalence rate in the population described by the latent distribution, $f(\phi)$, for the condition that ϕ exceeds ϕ^* is naturally defined as

$$P\{\phi \geq \phi^*\} = \int_{\phi^*}^{\infty} f(\phi)d\phi. \tag{15}$$

In addition to this overall prevalence rate, the prevalence in a subgroup of household indicated by $\mathbf{G} = g$ is given by

$$P\{\phi \geq \phi^* \mid \mathbf{G} = g\} = \int_{\phi^*}^{\infty} f(\phi \mid \mathbf{G} = g)d\phi. \tag{16}$$

However, the practice of USDA is to set the cut points on the scale of the manifest variables, that is, on the number of affirmations of the HFSSM questions. What is the connection between the proportions of households that affirm some number of the HFSSM questions and the prevalence rate defined in equation (15) or (16)? To answer this, let A denote the number of the dichotomized HFSSM questions affirmed by a household, then the percentage of households that affirm x or more of the questions is

$$P\{\mathbf{A} \geq x\} = \int_{-\infty}^{\infty} P\{\mathbf{A} \geq x \mid \phi\}f(\phi)d\phi. \tag{17}$$

Examples of subgroup prevalence rates are found in Table 2 of Nord et al. (2004). If the interest is on the percentage of households in a subgroup that affirm x or more of the questions, then, assuming no DIF, equation (17) is modified, as in equation (16), that is,

$$P\{A \geq x \mid G = g\} = \int_{-\infty}^{\infty} P\{A \geq x \mid \phi\} f(\phi \mid G = g) d\phi. \tag{18}$$

Equations (17) and (18) express the probabilities, $P\{A \geq x\}$ and $P\{A \geq x \mid G = g\}$, as the average of the conditional probability $P\{A \geq x \mid \phi\}$ in a standard way. For any plausible IRT model, $P\{A \geq x \mid \phi\}$ is an increasing function of ϕ, ranging from a small value for low values of ϕ to nearly 1 for large values of ϕ.

If, for some value of x, $P\{A \geq x \mid \phi\}$ were a step function, that was zero to the left of ϕ^* and 1 to the right of ϕ^*, then the prevalence rate in equation (15) and the affirmation rate in equation (17) would be equal. However, for any value of x, $P\{A \geq x \mid \phi\}$ is far from a step function, due to the inherent measurement error between the latent and manifest variables. It is possible that, for an appropriate choice of x, the parts of the function, $P\{A \geq x \mid \phi\}$, above and below the cut point, ϕ^*, would "balance," but this would have to be investigated in each case and for equation (18) could depend on the value of g. The difference between equations (15) and (17) is the bias arising from the use of a cut point based on the manifest variables and the use of one defined on the latent scale. This bias was addressed in a way in Nord (1999). How well he was able to investigate this bias is not clear to us due to the complexity of the task.

The form assumed for the latent distribution, $f(\phi)$, can make a difference in the estimated prevalence of food insecurity. This can be studied to some extent by trying out different assumptions and seeing what effect they have. As the number of manifest variables increases, the effects of different assumptions about the latent distribution grow less, but in the case of food insecurity the number of manifest variables is too small for this to be assumed.

Johnson (2005) describes several approaches to estimating prevalence rates that are defined by cut points along the ϕ-scale and of the form in equation (15) rather than equation (17). These methods avoid the biases mentioned above and apply to either overall or subgroup-specific prevalence rates. An example of the bias in prevalence estimates that arises from the failure to condition appropriately on the subgroup is given in Mislevy et al. (1992) for an educational testing application.

The Consequences of Measurement Error

As discussed earlier, the latent posterior distribution in equation (11), $f(\phi \mid X_1 = x_1, \ldots, X_p = x_p)$, summarizes all that is known about the latent variable from the values of the manifest variables. The fact that there is measurement error in the connection between the manifest variables and the latent variable in any latent variable model results in the latent posterior distributions being spread over a range of values along the ϕ-scale, rather than being concentrated on a single point along this scale. In the special circumstances in which many manifest variables are all strongly connected to the latent variable, the estimated posterior distribution is strongly peaked over a single value, so that it then makes sense to "estimate ϕ" by, for example, the unconditional maximum likelihood approach (Haberman, 1977; Holland, 1990). This situation often happens in educational testing applications of IRT models, in which the tests may comprise 40 to 100 test items.

However, in the case of the dichotomized HFSSM questions on the CPS, there are relatively few manifest variables on which to base our knowledge of ϕ for a given individual—at most 10 for households without children and 18 for those with children. In this circumstance, the estimated posterior distributions are not highly peaked over a single value of ϕ and spread over a range of values.

Johnson (2004, p. 23) gives a graph of two estimated posterior distributions that correspond to two different patterns of responses to the food insecurity questions. Johnson's graphs indicate that the estimated posterior distributions have substantial standard deviations, as one would expect, because of the small number of items. In addition, the two posterior distributions almost completely overlapped. Thus, measurement error is a significant aspect of the measurement of food insecurity by USDA.

An important consequence of this type of measurement error concerns the intuitively plausible use of the distribution of estimated values of the latent variable as a proxy of the latent distribution. These estimated values of the latent variable are a side benefit of the unconditional maximum likelihood method of estimating the item parameters. However, when the effect of measurement error is large, as it is in the case of food insecurity measurement, the distribution of the estimated values of ϕ across the sampled households does not form an unbiased estimate of the latent distribution.

As mentioned earlier, an additional issue relevant to measurement error is that when the number of manifest variables is relatively small, the form assumed for the latent distribution can affect the estimated latent posterior distribution and through that estimates of prevalence. Thus, even the form of the latent posterior distribution is somewhat uncertain when the number of items is small.

Stability of Scales over Time

The food insecurity scales were defined in the late 1990s. Unlike NAEP, the questions on the HFSSM remain the same year after year. Nevertheless, Ohls et al. (1999) discuss some indications that the scales were somewhat different for some of the years investigated. The source of this variability remains uncertain and could be due to several sources, for example, the poor fit of the Rasch model, technical aspects of the data collection, and different interpretations of the HFSSM questions in different years. The possibility that scale drift might occur should be examined on an ongoing manner to the extent possible. Methods of detecting differential item functioning may be used to investigate it with the data from different years being treated as the multiple groups.

BETTER MATCH BETWEEN THE MEASUREMENT MODEL AND THE DATA COLLECTED

The current approach to IRT modeling used by USDA is to create dichotomous/binary questions out of the several types of questions on the HFSSM, and then to use the Rasch model, which is designed for dichotomous questions. This approach has the potential of not using all of the information that is available in the battery of food insecurity questions. In addition, because two pairs of the questions are each two parts of a single question, the assumption of conditional independence in equation (5) is violated with unclear consequences. This section briefly outlines how this practice could be modified using ordered polytomous items.

Using Polytomous Items

In the current HFSSM, three types of questions are asked of either all households or households with children. First, there are dichotomous/binary questions with a yes/no response set. Second, there are questions with a trichotomous ordered response set (never, sometimes, often). Third, there are two-part questions that include a frequency follow-up to an initial question. The initial questions have a yes/no response set, and the frequency follow-up, if the initial answer is yes, has the trichotomous response set of (1–2 months, some months but not every month, or almost every month). For all households there is one dichotomous, three trichotomous, and four two-part questions. For households with children there is, in addition, one dichotomous, three trichotomous, and three two-part questions. Hence, for all households there are 8 different measures of food insecurity and for households with children there are an additional 7 measures, for a total of 15. These enumerations include only the questions that ask about "in the last 12 months" and omit those that ask about "in the last 30 days," because only the former are used to classify households and to estimate the prevalence rates.

The dichotomization of the trichotomous items is never/(sometimes or often); for the frequency follow-ups of the two-part questions, the dichotomy is 1–2 months/(some months but not every month or almost every month). The use of dichotomized responses carries with it the potential of a loss of information, but the degree of this loss is currently unknown for the data from USDA's food insecurity module (Ramsay, 1973). In addition, the approach used by USDA ignores the correlated nature of the dichotomized two-part questions.

It is possible to turn the responses to each two-part question into a single polytomous response set with four responses. These four responses are easily seen to be "no, yes in 1–2 months, yes in some months but not in every month, and yes in almost every month." These form an ordered set of responses that indicate an increased frequency, and therefore intensity of food insecurity.

There is a slight problem that missing data can create for this polytomization of the two-part questions. When there is an answer to the initial question of *yes* but, for some reason, there is no response to the follow-up question, the *yes* response can not be further pigeonholed. This can be handled in a variety of ways. For example, a conservative approach would be to assign a response of "yes in 1–2 months" to such cases. Other approaches might be considered as well.

Polytomous IRT Models for Polytomous Food Insecurity Questions

This section gives some details regarding an IRT model that is appropriate for polytomous food insecurity questions. The model developed here is one of several that allow for general ordered response categories rather than simply the binary case of endorse or not. It is an example of a threshold model described earlier for dichotomous items, and it includes that case as well. This more general approach can use all of the item response data that are currently available, including both the trichotomous questions and the two-part questions. In addition, it does not ignore the correlated aspect of the two-part questions.

The models described here are examples of *graded response models* (Samejima, 1969) and are specifically designed for ordered polytomous responses to the items. The panel considers them here because the measurement model is easy to understand, and it is sufficiently general for the currently collected food insecurity data. Different but related IRT models for ordered responses are given in Masters (1982) and Muraki (1992).

For concreteness, consider one of the trichotomous HFSSM questions with the possible responses of "never, sometimes, or often." Similar considerations apply to the above proposed way to make a single four-category item from a question with a frequency follow-up. The responses are ordered from least to most often. To specify a measurement model for this question, continue to assume that there is a latent variable, ϕ, that underlies a respondent's answer to food insecurity questions. The higher ϕ is, the more likely the household is to give a response of "often" for this question and the less likely a response of "never."

Because there are three possible responses, there will be two thresholds, b_1 and b_2, on the ϕ-scale (rather than the single threshold of the dichotomous case). Furthermore, since it is literally a "higher threshold," b_2 is greater than b_1. For the case of four ordered responses, there will be three thresholds to categorize the possible responses. The idea is the same as before for the dichotomous case. The location of ϕ with respect to these thresholds will determine the probability of the response of the household to the question. As in the dichotomous case, ϕ is perturbed by a statistically independent measurement error, V/a, and if $\phi + V/a$ is less than b_1, then the household's response is "never" to the question. If $\phi + V/a$ is between b_1 and b_2, then the household's response is "sometimes" to the question. If $\phi + V/a$ exceeds b_2, then the household's response is "often" to the question.

In the case of ordered responses, it is natural to define the item response function in terms of $P\{X \geq x \mid \phi\}$, the probability of a response equal to or exceeding x in intensity or frequency. Note that the notation $X \geq x$ makes sense because the possible responses are "ordered." In this case, X denotes the response of a household to the question, x denotes one of the values "sometimes" or "often," and ϕ is the household's value of the latent variable. When considering the probability, $P\{X \geq x \mid \phi\}$, it is useful to remember that the values of x that make sense to consider in a trichotomous example are the highest two, that is, "sometimes" and "often," because $P\{X \geq never \mid \phi\}$ always has the value 1.0 for any value of ϕ. As functions of ϕ, $P\{X \geq sometimes \mid \phi\}$ must be larger than $P\{X \geq often \mid \phi\}$ as ϕ varies, because "$X \geq sometimes$" means "$X = sometimes$ or $X = often$" and therefore it is more likely than "$X \geq often$."

The item response functions for this threshold model are given by

$$P\{X \geq sometimes \mid \phi\} = P\{\phi + V/a > b_1 \mid \phi\} \tag{19}$$

and

$$P\{X \geq often \mid \phi = P\{\phi + V/a > b_2 \mid \phi\} . \tag{20}$$

Rearranging terms in equation (19) and exploiting both the symmetry and independence of **V** shows that

$$P\{X \geq \textit{sometimes} \mid \phi\} = P\{V < a(\phi - b_1) \mid \phi\} = F_V(a(\phi - b_1)), \tag{21}$$

where $F_V(t)$ is the cdf of **V** as in equation (8). A similar formula holds for $P\{X \geq \textit{often} \mid \phi\}$ with b_2 replacing b_1.

The choice of $F_V(t)$ for **V** gives rise to different IRT models for this situation. If it is assumed that **V** has the normal distribution with mean 0 and variance 1, the result is the *Normal Ogive* graded response model, one of the earliest IRT models for ordered polytomous responses (Samejima, 1969). If **V** has a logistic distribution, the result is a logistic graded response model, of which the Rasch model is a dichotomous special case with $a = 1$. When a is allowed to vary across the items, the result is a graded version of the 2PL model in equation (7). The general family of models given by equation (19) or (20) describes a wide range of the common IRT models used in practice.

In the trichotomous case, the item parameters are a, b_1, and b_2. The item parameters affect the probability of the different responses—never, sometimes, and often—to the question for a household with the specified value of ϕ. The thresholds, b_1 and b_2, are examples of location parameters (akin to the single b parameter in the Rasch model for dichotomous responses), and they control how likely or unlikely the corresponding responses are for a given value of ϕ. The discrimination parameter, a, controls the amount of measurement error as it does in the dichotomous case.

The values of the item parameters depend on the nature of the question and the types of possible responses. For a question that asked if the respondent "worried about running out of money for food," one would expect that it does not have as large a b_1-value as does a question that asked if the respondent actually "did not eat because there was not enough money to do so." "Not eating because there was not enough money" is a more severe indicator of food insecurity than "worrying about food running out." For example, in the data published by the USDA for 2004, 16.6 percent of households affirmed the "worry" question, but only 3.1 percent affirmed the "not eating" question (Nord et al., 2005b). The model is consistent with a conception of food insecurity that assumes that households with larger values of ϕ are more likely to respond more intensely or frequently to a trichotomous or two-part indicator of food insecurity than those with lower values of ϕ.

The location of the b-parameters for a given question indicates the degree to which respondents do or do not endorse the question and at what level. More respondents affirm questions with lower b's and fewer respondents affirm questions with higher b's. This inequality reflects the way that the responses to the food insecurity questions are indicators of the underlying latent variable of food insecurity.

In order to complete the model for the joint distribution of responses to a set of food security questions, it is necessary to assume a latent distribution for ϕ. The simplest assumption is that ϕ has a Gaussian distribution with mean 0 and variance 1. However, due to the level of prior screening of households who are asked the FSS questions, Johnson (2004) suggested using a truncated Gaussian distribution, rather than the standard Gaussian distribution.

When there are several items being modeled, it is assumed that the measurement errors, V_i, are independent across the items, as in the dichotomous case. The independence of the measurement errors implies the conditional independence of the responses to questions given the latent variable ϕ, so that equation (5) holds.

As a final point, the panel observes that the type of IRT model just described allows frequency information that is currently collected to be more systematically examined than it is now. However, it does not address the measurement of the duration of spells of food insecurity. In order to address duration, more detailed data would need to be collected and more complicated time-dependent IRT models would need to be used to analyze them.

CONCLUSIONS AND RECOMMENDATIONS

Many issues have been raised in this chapter about the use of IRT models in the measurement of food insecurity and, in particular, the specific use of the Rasch model. These and related issues are summarized below and recommendations for improvement are presented.

1. Regarding food insecurity as an unobserved latent variable with the observed USDA survey questions regarded as a collection of manifest indicators of that latent variable is appropriate.
2. IRT models are appropriate for modeling survey responses that collect information on food insecurity of households.
3. Some empirical evidence suggests that the assumption of the Rasch model of equal item discrimination may not hold particularly well for the bank of survey items currently being used by USDA in its measurement of food insecurity.
4. The current practice in which responses to the food insecurity questions are dichotomized when, in fact, most of the questions have ordered polytomous response options may lead to a significant amount of information loss regarding that household's level of food

insecurity. This is especially of concern given the relatively few questions that define the food insecurity measure.

5. Several questions in the food insecurity supplement are follow-up questions asked only if "yes" is answered to a previous question. This structure should be taken into account in summarizing information about food insecurity. In particular, the responses to the stem item and the frequency follow-up question are not conditionally independent, an assumption made by the IRT model.

6. Under current practice, estimates of food insecurity at the household level ignore the large amounts of inherent uncertainty that exist due to its measurement by a small number of items. As a result, it is more appropriate to regard each household's value of ϕ as estimated by its posterior distribution of ϕ-values rather than by an estimated value that does not reflect its uncertainty. It is not appropriate to ignore this uncertainty when classifying households or estimating the prevalence of food insecurity, either overall or in subsets of households.

7. Currently USDA collects information on the intensity and frequency of food insecurity in U.S. households. The information does not address the measurement of the duration of spells of food insecurity, either overall or in subsets of households.

These conclusions lead to the following recommendations to improve the categorization of households into food insecurity levels:

Recommendation 5-1: USDA should consider more flexible alternatives to the dichotomous Rasch model, the latent variable model that underlies the current food insecurity classification scheme. The alternatives should reflect the types of data collected in the Food Security Supplement. Alternative models that should be formally compared include:

- Modeling ordered polytomous item responses by ordered polytomous rather than dichotomous item response functions.
- Treating items with frequency follow-up questions appropriately, for example, as a single ordered polytomous item rather than as two independent questions.
- Allowing the item discrimination parameters to differ from item to item when indicated by relevant data.

Recommendation 5-2: USDA should undertake the following additional analyses in the development of the underlying latent variable model:

- Fitting models that allow for different latent distributions for households with children and those without children and possibly other subgroups of respondents.
- Fitting models that allow for different item parameters for households with and without children for the questions that are appropriate for all households in order to study the possibility and effects of differential item functioning.
- Studying the stability of the measurement system over time, possibly using the methods of differential item functioning.

Recommendation 5-3: To implement the underlying latent variable model that results from the recommended research, USDA should develop a new classification system that reflects the measurement error inherent in latent variable models. This can be accomplished by classifying households probabilistically along the latent scale, as opposed to the current practice of deterministically using the observed number of affirmations. Furthermore, the new classification system should be more closely tied to the content and location of food insecurity items along the latent scale.

Recommendation 5-4: USDA should study the differences between the current classification system and the new system, possibly leading to a simple approximation to the new classification system for use in surveys and field studies.

Recommendation 5-5: USDA should consider collecting data on the duration of spells of food insecurity in addition to the currently measured intensity and frequency measures. Measures of frequency and duration spells may be used independently of the latent variable measuring food insecurity.

6

Survey Vehicles to Measure
Food Insecurity and Hunger

The U.S. Department of Agriculture (USDA) bases its annual report and estimates of the prevalence of food insecurity on data collected from the Food Security Supplement (FSS) to the Current Population Survey (CPS). The Household Food Security Survey Module (HFSSM) within the FSS, or a modification of it, is or has been used in several surveys. One of the main objectives of the annual food insecurity measure is to monitor the estimated prevalence of food insecurity and changes in the prevalence over time at the national and state levels for purposes of program policies, as well as the need for program development.

Attaching the FSS to the CPS for the official estimates has several advantages. The CPS is the largest of the major national sample surveys and comes closest to fulfilling one of the key goals for the project—national and state-level monitoring of food security. Data from the FSS in the CPS are released on a timely basis. There are, however, reasons to consider other surveys as a possible primary vehicle or to complement the FSS food insecurity data for research purposes. This chapter briefly reviews the key features of selected national surveys and assesses the relative advantages and disadvantages of each survey reviewed.

KEY FEATURES OF SELECTED SURVEYS[1]

Current Population Survey

The design of the CPS is described in Chapter 4 of this report. To summarize, the CPS, sponsored jointly by the Bureau of Labor Statistics and the Census Bureau, is a national household sample survey. Like most major national surveys, the CPS covers the civilian noninstitutionalized household population, including noninstitutionalized groups not living in conventional housing units and groups living in housing containing nine or more persons unrelated to the person in charge. Its design is a nationally and state-level representative sample, but estimates for small states have large confidence intervals. It has a very large sample size of 60,000 households (about 119,000 individual respondents in households, with about 31,000 under age 18) interviewed every month. In addition to its core content, a different supplement is fielded each month. The CPS has high response rates, and adding the FSS to it is relatively inexpensive and allows for timely production of annual prevalence data on food insecurity. Data are available from the Census Bureau within 3–4 months and released by USDA in published form to the public about a year after collection. The cost of the supplement to USDA is about $450,000 each year. This includes data collection, initial editing for confidentiality and weighting, incorporating household-level food security variables calculated from the initial data by USDA, preparation of documentation for the public use file, and purveying the public use data on CD-ROM and on the Census Bureau's Data FERRET system.[2]

As noted earlier, CPS uses a rotating sample design and technically is a panel survey because a household is in the survey for four months, then out for eight months, and then back in for four months. The sample unit is the household address and not the household, and for the reinterview period the original occupants of the address may have moved since the earlier interviews. One person is interviewed to obtain information for the entire household. The CPS is in the field every month, but only the Annual Social and Economic Supplement to the CPS in March and some rotation groups in February and April include detailed information on household income. The FSS is administered in December, at which time very little income information is collected. The CPS does not collect information on general health of the population or nutritional intake.

---

[1]Some of the information in this chapter is drawn from the background paper prepared for the panel by Haider (2005).

[2]Email communication with Mark Nord, Economic Research Service, USDA, on September 25, 2005.

National Health Interview Survey

The National Health Interview Survey (NHIS), conducted by the National Center for Health Statistics (NCHS), is a multipurpose health survey and the principal source of basic information on the health of the civilian noninstitutionalized household population of the United States. It is a cross-sectional, nationally representative sample with reliable estimates for the four defined geographic regions of the United States.

The NHIS consists of a core or basic module as well as variable supplements. The basic module consists of three components: the family core, the sample adult core, and the sample child core. The family core component collects information on everyone in the family, including household composition and sociodemographic characteristics; tracking information; information that matches administrative databases and basic indicators of health status; activity limitations, injuries, health insurance coverage, and access to, and use of, health care services. For each family in the NHIS, one sample adult and one sample child (if any children under age 18 are present) are randomly selected and information is collected on their health status, health care services, and behavior with the specific core questionnaires.

For the family core questionnaire of the basic module, all members of the household age 18 and over who are at home at the time of the interview are invited to participate and to respond for themselves. For children and adults not at home during the interview, information is provided by a knowledgeable adult residing in the household. For the sample adult questionnaire, one adult per family is randomly selected; this individual responds for himself or herself, unless physically or mentally unable to do so. Information for the sample child questionnaire is obtained from a knowledgeable adult residing in the household. Topical modules are fielded less frequently.

About 40,000 households are interviewed annually, including approximately 112,000 people of whom 29,000 are under age 18. The interviewed sample for 2004 consisted of 36,579 households, which yielded 94,000 persons in 37,466 families. The interviewed sample for the sample child component, by proxy response from a knowledgeable adult in the family, was 12,424 children under age 18. Interviewing is conducted continually, so that seasonality can be studied from any systematic month-to-month variation over the year and the effects of seasonality can be averaged out.

NHIS is not currently designed to provide state-level estimates. However, black and Hispanic populations are oversampled now to allow more precise estimation of health in these growing minority populations.

A possible problem is that the NHIS is faced with periodic budgetary shortfalls, as occurred in 2002, 2003, and 2004. As a result, NCHS reduced the size of the NHIS sample in these years. Since grouping the sample cuts into consecutive weeks yields the greatest monetary savings, the NHIS sample was cut in three consecutive weeks in May 2004. For 2003 and 2002, the NHIS sample was cut in three consecutive weeks in April 2003 and two consecutive weeks in December 2002. Thus no new interviews were conducted during 3 of the normal 50 weeks of interviewing.

The data collected in the NHIS contain extensive measures of health and well-being of the respondents. Supplementary modules on different health-related topics are added by sponsoring agencies, and they vary from year to year. Some recent and planned examples are:

- 2000: Cancer risk factors (National Cancer Institute); also 1987 and 1992
- 2001: Healthy People 2010; child mental health (National Institute of Mental Health)
- 2002: Healthy People 2010; complementary and alternative medicine (National Center for Complementary and Alternative Medicine); child mental health (National Institute of Mental Health)
- 2003: Healthy People 2010; child mental health (National Institute of Mental Health)
- 2004: Child mental health (National Institute of Mental Health); cell phone questions
- 2005: Cancer risk factors (National Cancer Institute); child mental health (National Institute of Mental Health); cell phone questions
- 2006: Diabetes (National Institute of Diabetes and Digestive and Kidney Diseases); child mental health (National Institute of Mental Health)
- 2007: Complementary and alternative medicine (National Center for Complementary and Alternative Medicine)

NHIS also includes questions about participation in the most common assistance programs. Although it asks about household income, the income module is brief and does not ask about food expenditures.

Data from the supplements are released in the same files as the core data and are released at the same time as data from the basic module. The timing of data release has improved noticeably in the past few years, and NCHS is working to improve it further. Quarterly estimates for 15 key health indicators are released on its website about 6 months after data collection. In general, other data are released in published form 14–18 months after the end of the data collection year.

National Health and Nutrition Examination Survey

The National Health and Nutrition Examination Survey (NHANES), conducted by NCHS, collects detailed anthropometrical, medical, and nutritional information on all sample persons. These data are valuable for understanding the links between food insecurity and health and between food insecurity and diet. Food security and sufficiency have been measured in NHANES since NHANES III. From 1999 to 2001 the food sufficiency question was expanded to include food adequacy. The question was dropped beginning in 2002. NHANES has included the 18-item household food security survey module in the family interview part of the household interview since 1999. Individually referenced food security questions (7 for participants age 16 and older, 6 for participants younger than age 16) were added to the postdietary recall component of the examination section and were released with the 2001–2002 food security data release. NHANES continues to collect both the household and individual-level food security data in collaboration with USDA.

Beginning in 2000, NHANES included questions about the individual-level hunger of participants age 16 and over and of a proxy regarding children under age 12. Beginning in 2005, NHANES will ask 12–16-year-olds these questions. These questions ask about the individual's experience, in contrast to the household-level questions, which just ask about anyone in the household. These individual-level responses can then be assessed in relation to individual measures from other examination components.

NHANES collects an appreciable amount of dietary information. It uses a dietary recall method to collect information on the complete dietary intake for two nonconsecutive 24-hour periods, and supplemental questions are asked about the intake of infrequently consumed food for the past 30 days. In addition, it collects self-reports about health conditions and includes a detailed medical examination including blood analysis.

Similar to NHIS, NHANES also includes questions about participation in the most common assistance programs. Although it asks about household income, the income module is brief. The sampling for NHANES is complicated. Although it selects individuals in the same household at a higher probability than individuals in different households, individuals in a household are still subject to selection in order for NHANES to achieve its complex oversampling scheme. It uses a household respondent for various household characteristics but then interviews, and gives a medical examination to, each sampled person.

NHANES was initially conducted on a periodic basis, but since 1999 it is conducted on a continuing basis. The NHANES samples about 5,000 households annually, so it is not large enough for annual estimates or for subgroup or state analysis. It groups two years of data to develop national estimates, and these two-year blocks of data are available for research purposes.

Survey of Income and Program Participation

The Survey of Income and Program Participation (SIPP), conducted by the Census Bureau, is a nationally representative household sample survey of the noninstitutional population. The sample is drawn to be representative at the state level, but estimates for most states have large confidence intervals. The survey design is a continuous series of national panels, with sample size ranging from about 14,000 to 37,000 interviewed households. Although the sampling unit is the household, SIPP attempts to interview all household members age 15 and older by self-response. SIPP is designed as a panel survey of all individuals in a household, and therefore movers are followed. A new SIPP panel was fielded every three to four years over the past decade, and each panel is reinterviewed every four months. The last complete SIPP panel began in 2001 and interviewed nearly 37,000 households in a panel design of 9 waves spaced 4 months apart. The current SIPP panel, started in 2004, has approximately 43,000 households interviewed at Wave 1 (110,700 individuals of whom 29,700 are under age 18). It is scheduled for 12 waves.

SIPP is divided into core content, which is collected during each wave, and topical content, which is collected only during certain waves. SIPP collects detailed social and economic information, including program participation. It also collects some self-reported health measures, such as limitations in activities of daily living and self-assessed health status. Besides some health information, it queries income sources so that household income can be constructed. In principle, then, SIPP can be used to construct a household-level variable, such as income or wealth. However, SIPP does not obtain information on food expenditures. Low-income households are oversampled, resulting in about an 11 percent increase in the number of low-income households compared with what would be without oversampling.[3] SIPP has included a subset of six food insecurity questions (but not the standard six-item set) in the adult well-being module once during each panel beginning in 1998.

Because of funding restrictions, the sample size and design of SIPP has changed often in the past. Whether it will retain a stable design and sample size is not known, but the changing design is a feature of SIPP that makes it less attractive than, say, the CPS. Moreover, the content of SIPP is decided

---

[3]Email communications on June 2, 2005, with Kathleen P. Creighton, U.S. Census Bureau.

by an interagency committee, which has to review an agency's request before questions can be added to the survey. The survey has a lengthy processing cycle. The period between data collection and data availability can be one to two years.

RELATIVE ADVANTAGES AND DISADVANTAGES

Because food insecurity and hunger as now measured are relatively rare events, fairly large sample sizes are required to estimate accurately their prevalence in subpopulations of interest.

The CPS December interview has been used for the past few years as the vehicle for the Food Security Supplement. An important advantage of CPS is its widespread acceptance as an authoritative source of statistical information. A second advantage is its sample size of about 50–60,000 households and its state-level representative sample. Thus one can get reliable estimates of prevalence in subpopulations of interest. A third advantage is the timeliness of the reporting of data. CPS does, however, have a number of important disadvantages. It is a survey of households; only the respondent is surveyed, so that any reports about the hunger of the individuals in the household are by proxy. A second disadvantage is that detailed income information is collected in the CPS only in the Annual Social and Economic Supplement. Household income may have changed, and that very change may have given rise to food insecurity. Household composition may have changed, resulting in the loss of income or the gain of income. As with most cross-sectional household surveys, CPS is based on dwelling units, so that subpopulations that move often would be less likely to be interviewed in both December and the Social and Economic Supplement. No health information is obtained in the CPS.

For reasons of sample size, and not necessarily the content, the panel concludes that NHANES is not an appropriate primary vehicle for monitoring the prevalence of food insecurity in the population at the national and state levels. *It could be an important vehicle for a research program for understanding the relationship between food insecurity and hunger and indicators of inadequate nutrition and other health characteristics.* It could also be an appropriate vehicle for the new research recommended in Chapter 3 on aspects of deprivation, alienation, and family and social interactions. As indicated in the previous section, USDA has been collaborating with NHANES in collecting data on the HFSSM and has started analyzing those data.

Recommendation 6-1: USDA should continue to collaborate with the National Center for Health Statistics to use the National Health and Nutrition Examination Survey to conduct research on methods

of measuring household food insecurity and individual hunger and the consequences for nutritional intake and other relevant health measures.

In Chapter 3 the panel concluded that food insecurity is a household-level concept. It is rooted in the lack of economic resources in a household. Hunger is distinct from food insecurity and is an individual-level concept. Both measures are important, and to measure both concepts, the panel recommends that USDA should undertake a research program on how best to measure individual hunger and other important consequences of food insecurity. The CPS cannot now be used to measure hunger in the population because it interviews only the household respondent. While the CPS can ask the household respondent about his or her experience with hunger, no one respondent is likely to be representative of all adults in the household.

The NHIS interviews a randomly chosen adult who can be asked about his or her experience with hunger. NHIS is based on a probability sample; one can use the responses to estimate the prevalence of hunger in the adult population. With the randomly chosen child, one can estimate the prevalence of hunger among children. Similarly, the consequences to individuals of food insecurity or hunger, such as health problems and social exclusion or alienation, can be ascertained from the interviews. Another major positive aspect of the NHIS is its extensive measures of health, which could be linked to food insecurity or hunger.

Recall bias is also a concern about the current reliance on the December CPS. Asking a respondent to recall instances of food insecurity over a 12-month period is likely to produce recall bias, in which the respondent overweights the current situation. This bias is probably exacerbated by the measurement of food insecurity in the December CPS because December is not representative of the experience over the whole year. Although the FSS has probe questions and questions about a 30-day recall, the NHIS collects data across the entire year (which increases the cost of the supplement to USDA) and also has the capability to estimate the frequency and duration of food insecurity. Ultimately, USDA would need to weigh comparative costs and benefits, timeliness, and the ability to include the supplement on a regular basis in the NHIS, among other issues.

> **Recommendation 6-2: USDA should carefully review the strengths and weaknesses of the National Health Interview Survey in relation to the Current Population Survey in order to determine the best possible survey vehicle for the Food Security Supplement at a future date. In the meantime, the Food Security Supplement should continue to be conducted in the Current Population Survey.**

The four-month intervals between SIPP interviews make it well suited to ask about intervals of food insecurity and/or hunger over both short intervals such as last week or last month, or over the last four months. Because SIPP measures contemporaneous income, analysis can find the relationships between income and food insecurity and/or hunger. This could be particularly important to find the direct effects of income loss via, say, unemployment or a health shock and to separate out short-term from long-term food insecurity.

CPS is limited in its ability to obtain information that would permit the scientific study of the antecedents and consequences of food insecurity and hunger. Such studies require panel data, such as in SIPP. Panel data would provide an additional benefit to USDA. They would permit a better assessment of the success of such programs as food stamps, because analysts could study the dynamics of economic need, food insecurity, and subsequent relief due to food stamps.

The panel recognizes that in an era of budgetary constraints it would be difficult for USDA to accomplish all the goals of reliably measuring food insecurity prevalence, hunger and their dynamics. But we note that not all measures need to be collected at all times. In particular, USDA should explore the feasibility of a one-time multiwave study using SIPP to examine the dynamic relationships among income, food insecurity, and hunger. Especially useful would be a study of the persistence of food insecurity and hunger, including its prevalence and frequency. The four-month interviewing cycle of SIPP should provide much more accurate information than the CPS. In addition, data from the shorter recall period could be compared with those from a 12-month recall period to study any bias in the current 12-month recall design.

Recommendation 6-3: USDA should explore the feasibility of funding a one-time panel study, preferably using the Survey of Income and Program Participation, to establish the relationship between household food insecurity and individual hunger and how they co-evolve with income and health.

7

Applicability of Food Insecurity Outcomes for Assessment of Program Performance

T his chapter examines the use of the annual prevalence estimates of food insecurity in the United States for assessing the performance of the U.S. Department of Agriculture's (USDA) food assistance programs in accordance with the Government Performance and Results Act of 1993. Since specific recommendations and suggestions discussed earlier in the report may lead to changes in the current system, the objective in this chapter is to consider this issue from a broad perspective.

FOOD SECURITY AS A MEASURE OF PROGRAM PERFORMANCE

The Government Performance and Results Act of 1993 (GPRA) requires government agencies to account for the intended results of their activities. Although it provided for strategic planning and managerial accountability, the primary goal of the law was to force a shift in government programs away from process goals (like the number of grants made) to measuring outcomes—whether the intended results of the program were achieved. Federal government agencies were required to "improve program effectiveness and public accountability by promoting a new focus on results, service quality, and customer satisfaction." The law requires that specific performance goals be established and that annual measurement of these output goals be undertaken to determine the success or failure of the program (Government Performance and Results Act, 1993, Section 2. Findings and Purposes, no. 2).

USDA's Food and Nutrition Service (FNS) is the agency responsible for the major food assistance programs, including the Food Stamp Program, the Special Supplemental Nutrition Program for Women, Infants, and Children, and the National School Lunch program. The 2000–2005 strategic plan for FNS states as a goal for the agency, in delivering the food assistance programs, to reduce the prevalence of "food insecurity with hunger" among households with incomes under 130 percent of the federal poverty standard.[1] Currently, FNS uses trends in the prevalence of food insecurity with hunger based on the food insecurity module included in the annual Food Security Supplement (FSS) to the Current Population Survey (CPS) as a measure of its annual performance. The panel was asked to comment on the applicability of these data for this purpose.

The mission of FNS is "to increase food security and reduce hunger in partnership with cooperating organizations by providing children and low-income people access to food, a healthful diet, and nutrition education in a manner that supports American agriculture and inspires public confidence." Legislative language authorizing the Food Stamp Program in the 1977 Food Stamp Act explicitly mentions the alleviation of hunger as a program goal. Therefore, it would seem reasonable to use trends in the prevalence of food insecurity and hunger in low-income populations as performance indicators to assess whether FNS is fulfilling its mission.

Estimates of the prevalence of food insecurity or hunger—when properly designed and measured—can be helpful in determining whether a population is experiencing more or less food insecurity over time. In principle such estimates could also serve as an important tool for identifying trends and levels in food insecurity for specific subgroups of households, such as the elderly, and for different geographic areas, such as rural areas and specific regions or states. Such monitoring efforts are important because they can help to identify where additional assistance may be needed, or where it can be reduced.

The methodological and conceptual issues regarding the measurement of food insecurity and hunger have already been reviewed in this report. There are many actual and possible limitations of the CPS and all other national household surveys in sampling the lowest income households in American society. As explained in Chapter 1, as with most national household population surveys, the CPS routinely excludes people who are institutionalized and those homeless people who cannot be found in households or other living quarters visited during household surveys. The panel recognizes the likelihood of relatively high rates of food insecurity among homeless, and the resulting negative bias resulting from their exclusion. At the

---

[1]See Wilde (2004) for a discussion of the use of the food insecurity measure for other performance assessment tools—for example, use with the Healthy People 2010 goals.

same time, it has serious questions about the operational and methodological issues. The panel concludes that until better methods to survey the homeless are developed, continuing to limit the target population to the household population seems appropriate. The panel, however, urges USDA to undertake research as part of its long-term agenda, leading to obtaining estimates of food insecurity in periodic or a one-time survey to get a sense of the negative bias of excluding this population in the household survey.

However, even an appropriate measure of food insecurity or hunger using appropriate samples would not be a definitive performance indicator of food assistance programs because their performance is only one of many factors that result in food insecurity. Consequently, changes in food insecurity could be due to many factors other than the performance of the food safety net. For example, if food insecurity declined because the price of food declined, the decline in food insecurity may not indicate better performance of the food safety net, because these programs would not have been responsible for all or part of the change. Conversely, if the reverse were to occur—that is, if food prices were to rise steeply or household income were to fall—the result might be an increase in the number of food-insecure households. But this, too, would not be the result of a decline in the performance of the food safety net. Developments like these could result in errors in assessing the performance of such programs as food stamps that are intended to reduce food insecurity and hunger.

USDA staff and colleagues have studied issues of this kind in a number of ways. They have compared the food security data with information collected in other surveys, such as the Survey of Income and Program Participation (SIPP) and the National Health and Nutrition Examination Survey (NHANES). They are currently using the panel feature of the CPS to look at the food insecurity of households as they approach the beginning of a food stamp spell—a period of one or more months during which a household received food stamps every month. Analysis is still in progress. A recent paper by Wilde and Nord (2005) used the food security data collected in 2002 and 2003 to estimate the effect of Food Stamp Program participation on food security. They used CPS Food Security Supplement data for December 2001 and December 2002 to determine the change in food insecurity with hunger status for food stamp participants who were in the survey both in December 2001 and in December 2002. They found that only 41 percent of the food-insecure households with hunger in 2001 had become food secure in 2002. They concluded that "it appears that unobserved hardships strike from time to time, with large effects on both program participation and food security. These hardships are sufficiently severe to swamp the presumably beneficial direct effect of food stamps on food security" (Wilde and Nord, 2005, pp. 430–431).

To better attribute changes in food insecurity or hunger to the food safety net would require that trends in these outcomes be supplemented with estimates of changes in need. FNS could annually analyze trends in the prevalence of food insecurity, taking into account economic and other changes that might affect that trend. Ideally, the food safety net would be able to quickly respond to changes in need so as to prevent an increase in food insecurity when need changes. Realistically the food safety net probably cannot fully do this. For example, food stamps cannot help people when food prices increase unless their income declines to make them eligible for food stamps.

The GPRA intended that agencies develop assessment measures that would allow them to assess the efficiency of their programs as well as their effectiveness. Trends in food insecurity or hunger tell us little about changes in the efficiency of FNS. Imagine that changes in food prices or other factors reduce the need for the food safety net, so that food insecurity declines among the poor. The trend in the prevalence of food insecurity would suggest that the food safety net was doing a better job at meeting the need for food assistance. But it is conceivable that prices could drop so much that the poor actually may need less than the safety net offered.

The GPRA emphasis on performance indicators requires agencies to define their goals in quantitative standards related to the purposes of the programs. In this respect, it is important to distinguish performance measurement from program evaluation. Definitions of program measurement and program evaluation promulgated by the U.S. General Accounting Office (1998) identify performance measurement as the ongoing monitoring and reporting of program progress toward preestablished goals. Program evaluations, by contrast, are individual systematic studies conducted periodically or on an ad hoc basis to assess how well a program is working. Program evaluation examines achievement of program objectives in the context of other factors in the program performance that may impede or contribute to its success.

The difficulty facing FNS (and other government agencies)—once having found a way to quantify a program outcome—is that they often must take account of other factors that might affect the need for the program. This can be extremely difficult. For the food safety net programs, this suggests that simple annual trends in food insecurity and hunger would have to be supplemented with estimates of changes in need as well as other possible economic changes that could affect the program participants. It may be difficult, sometimes impossible, to come up with reliable assessment indicators that fulfill the broad intent of GPRA. For example, to assess the effectiveness of the Food Stamp Program, FNS each year would have to analyze trends in the prevalence of food insecurity and hunger among the poor, taking account also of changes in income, food prices and scarcities, and other relevant matters that might affect that trend. The staff of FNS has already done a considerable amount of micro-level research as they have worked to understand the benefits achieved by the Food Stamp Program. They should continue this research.

The panel concludes that relying on trends in prevalence estimates of food insecurity as a sole indicator of program result is inappropriate. To assess programmatic results, better understanding is needed of the transitions into and out of poverty made by low-income households and the kind of unexpected changes that frequently bring about alterations—for good or bad—in households participating in food assistance programs.

8

Closing Remarks

During the two years this panel has met, it has reviewed a large number of research reports and journal articles, government studies, and relevant internal unpublished papers provided by the U.S. Department of Agriculture (USDA). Experts in the field participated in a large workshop organized by the panel and addressed the panel during its meetings. In this report, the panel has examined the issues that surfaced in the workshop, by USDA, by the panel's own discussions, and in other ways.

The panel is impressed with the extensive research thus far undertaken and with the continuing program of research carried out by USDA. It urges USDA to continue the research program and makes recommendations for its direction in the future.

The panel concludes that the measurement of food insecurity and of hunger is important. The recommendations in the preceding chapters are intended to improve this measurement, so that policy makers and the public can be better informed. Toward this end, it has recommended research efforts that should lead to improved concepts, definitions, and measurement of food insecurity and hunger in this country. The panel has provided a detailed discussion of the analytical methods used by USDA and made recommendations for modifying the model currently used by USDA and recommends further research that would lead to improved accuracy of the insecurity scale. The panel has also recommended research and testing to consider the strengths and weaknesses of the major national surveys in relation to the Current Population Survey in order to determine the best survey vehicle for the Food Security Supplement. At the same time it recognizes that such research will take time.

Overall, the panel concludes that the highest research priority is to develop a clear conceptual definition of hunger and to determine how best to implement the new definition, to study the manner in which item response theory models are applied to the data so that the classification structure better reflects the data collected in the FSS, and to take advantage of new developments in cognitive questionnaire design in evaluation and testing of the questions asked.

The panel commends USDA and its collaborating agencies for the careful and extensive work that went into the development of a standard food security measure. It is unusual for an agency to undertake such comprehensive research prior to the start of a survey, and the panel has been very impressed by much that has been done. The panel further recognizes USDA's continuing efforts to study, evaluate, and improve the measure. The additional research proposed by the panel builds on some of the work that USDA has already done. In addition, over the 10 years since the survey was launched there have been advances in survey design, questionnaire development, and modeling. The panel concludes that some of the new research suggested in the report could be especially useful, particularly in the key areas identified in the report.

Finally, the panel hopes that the points made in this report contribute toward development of a revised, efficient, and more cost-effective system for monitoring the prevalence of food insecurity in the United States. In addition, the suggestions for new research should help to link a household's food insecurity status with research to answer the important question about the broader health, socioeconomic, and psychological consequences of food insecurity in the United States.

References

Adams, E.J., Grummer-Strawn, L., and Chavez, G. (2003). Food security is associated with increased risk of obesity in California women. *Journal of Nutrition, 133*, 1070–1074.

Adams, R., and Wu, M. (Eds.). (2002). Pisa 2000 technical report. Paris: Organization for Economic Co-Operation and Development. (http://www.pisa.oecd.org/dataoecd/53/19/33688233.pdf).

Alaimo, K., and Froelich, A. (2004). *Alternative construction of a food security and hunger measure from 1995 Current Population Survey Food Security Supplement data.* Paper presented at the Workshop on the Measurement of Food Insecurity and Hunger, July 15, 2004. Panel to Review the U.S. Department of Agriculture's Measurement of Food Insecurity and Hunger. (http://www7.nationalacademies.org/cnstat/froelich_alaimo_paper.pdf).

Alaimo, K., Olson, C.M., and Frongillo, E.A. (1999). Importance of cognitive testing for survey items: An example from food security questionnaires. *Journal of Nutrition Education, 31*(5), 269–275.

Alaimo, K., Olson, C.M., and Frongillo, E.A. (2001a). Low family income and food insufficiency in relation to overweight in U.S. children—Is there a paradox? *Archives of Pediatrics and Adolescent Medicine, 155*, 1161–1167.

Alaimo, K., Olson, C.M., and Frongillo, E.A. (2001b). Food insufficiency and American school-aged children's cognitive, academic and psychosocial development. *Pediatrics, 108*, 44–53.

Alaimo, K., Olson, C., Frongillo, E., and Briefel, R. (2001c). Food insufficiency, poverty, and health in U.S. pre-school and school-age children. *American Journal of Public Health, 91*, 781–786.

Alaimo, K., Olson, C.M., and Frongillo E.A. (2002). Family food insufficiency, but not low family income, is positively associated with dysthymia and suicide symptoms in adolescents. *Journal of Nutrition, 132*, 719–725.

Andersen, E.B. (1980). *Discrete statistical models with social science applications.* Amsterdam: North-Holland.

Anderson, S.A. (1990). Core indicators of nutritional state for difficult-to-sample populations. *Journal of Nutrition, 120,* 1557–1600.

Anderson, T.W. (1954). On estimation of parameters in latent structure analysis. *Psychometrika, 19,* 1–10.

Andrews, M.S., Bickel, G., and Carlson, S. (1998). Household food security in the United States in 1995: Results from the Food Security Measurement Project. *Family Economics and Nutrition Review, 11*(1 and 2), 17–28.

Andrews, M.S., and Prell, M.A. (2001a). *Second food security measurement and research conference, volume I: Proceedings.* Food Assistance and Nutrition Research Report Number 11-1. Washington, DC: U.S. Department of Agriculture, Economic Research Service. (http://www.ers.usda.gov/publications/Fanrr11-1/).

Andrews, M.S., and Prell, M.A. (2001b). *Second food security measurement and research conference, volume II: Papers.* Food Assistance and Nutrition Research Report Number 11-2. Washington, DC: U.S. Department of Agriculture, Economic Research Service. (http://www.ers.usda.gov/publications/Fanrr11-2/).

Ashiabi, G. (2005). Household food insecurity and children's school engagement. *Journal of Children and Poverty, 11,* 3–17.

Bartholomew, D.J. (1987). *Latent variable models and factor analysis.* London: Charles Griffin.

Bavier, R. (2004). *Critique presented at the first meeting of the Panel to Review the USDA's Measurement of Food Insecurity and Hunger.* March 30. Office of Management and Budget.

Bickel, G., Nord, M., Price, C., Hamilton, W., and Cook, J. (2000). *Guide to measuring household food security, revised 2000.* Alexandria, VA: U.S. Department of Agriculture, Food and Nutrition Service.

Birdsall, T.G. (1955). The theory of signal detectability. In H. Quastler (Ed.), *Information theory in psychology* (pp. 391–402). Glencoe, IL: Free Press.

Birnbaum, A. (1968) Some latent trait models and their use in inferring an examinee's ability. In F.M. Lord and M.R. Novick (Eds.), *Statistical theories of mental test scores* (pp. 397–479). Reading, MA: Addison-Wesley.

Bock, R.D., and Lieberman, M. (1970). Fitting a response model for n dichotomously scored items. *Psychometrika 35,* 179–197.

Bohrnstedt, G.W. (1983). Measurement. In P.H. Rossi, J.D. Wright, and A.B. Anderson (Eds.), *Handbook of survey research* (pp. 70–114). Orlando, FL: Academic Press.

Bound, J., Brown, C., and Mathiowetz, N. (2001). Measurement error in survey data. In J. Heckman and E. Leamer (Eds.), *Handbook of econometrics,* Vol. 5. Amsterdam: North-Holland.

Bradlow, E.T., Wainer, H., and Wang, X. (1999). A Bayesian random effects model for testlets. *Psychometrika, 64,* 153–168.

Casey, P.H., Szeto, K., Lensing, S., Bogle, M., and Weber, J. (2001). Children in food-insufficient, low-income families—Prevalence, health, and nutritional status. *Archives of Pediatrics and Adolescent Medicine, 155,* 508–514.

Cohen, B., Parry, J., and Yang, K. (2002). *Household food insecurity in the United States, 1998–1999: Detailed statistical report.* E-FAN-02-011, prepared by IQ Solutions and USDA. Washington, DC: U.S. Department of Agriculture, Economic Research Service. (http://www.ers.usda.gov/publications/efan02011/).

Cohen, B., Ohls, J., Andrews, M., Ponza, M., Moreno, L., Zambrowski, A., and Cohen, R. (1999). *Food stamp participants' food security and nutrient availability.* Alexandria, VA: U.S. Department of Agriculture, Food and Nutrition Service. (http://www.fns.usda.gov/oane/MENU/Published/NutritionEducation/Files/nutrient.pdf).

Cook, J.T., Frank, D.A., Berkowitz, C., Black, M.M., Casey, P.H., Cutts, D.B., Meyers, A.F., Zaldivar, N., Skalicky, A., Levenson, S., et al. (2004). Food insecurity is associated with adverse health outcomes among human infants and toddlers. *Journal of Nutrition, 134*, 1432–1438.

de Onis, M., Blössner, M., Borghi, E., Frongillo, E.A., and Morris, R. (2004). Meeting international goals in child malnutrition. *Journal of the American Medical Association, 291*, 2600–2606.

Derrickson, J.P., Sakai, J.M., and Anderson, S.A. (2001). Interpretations of the "balanced meal" household food security indicator. *Journal of Nutrition Education, 33*(3), 155–160.

Dunifon, R., and Kowaleski-Jones, L. (2003). The influences of participation in the National School Lunch Program and food insecurity on child well-being. *Social Service Review, 77*, 72–92.

Dykema, J., and Schaeffer, N. (2005). *Cognitive aspects of the questions used to measure food insecurity and hunger.* Background paper prepared for the Panel to Review the U.S. Department of Agriculture's Measurement of Food Insecurity and Hunger. (http://www.nationalacademies/CNSTAT/).

Eisinger, P. (1996). Toward a national hunger count. *Journal of Social Service Review*, June, 214–234.

Eisinger, P. (1998). *Toward an end to hunger in America.* Washington, DC: Brookings Institution Press.

Federal Register. (1993). *Ten year plan for the National Nutrition Monitoring and Related Research Program. Part II.* 58:32752–32806. Washington, DC: U.S. Department of Health and Human Services and U.S. Department of Agriculture.

Froelich, A.G. (2002, November 21). *Dimensionality of the USDA Food Security Index*, Department of Statistics, Iowa State University.

Frongillo, E.A., Chowdhury, N., Ekström, E.C., and Naved, R.T. (2003). Understanding the experience of household food insecurity in rural Bangladesh leads to a measure different from that used in other countries. *Journal of Nutrition, 133*, 4158–4162.

Frongillo, E.A., and Horan, C.M. (2004). Hunger, food insecurity, and aging. *Generations, 28*(3), 62–63.

Frongillo, E.A., and Nanama, S. (2003). Development and validation of a questionnaire-based tool to measure rural household food insecurity in Burkina Faso. In *Proceedings of the international scientific symposium on measurement and assessment of food deprivation and undernutrition* (pp. 309–310). June 26–28, 2002. Rome: Food and Agriculture Organization of the United Nations.

Frongillo, E.A., Rauschenbach, B.S., Olson, C.M., Kendall, A., and Colmenares, A.G. (1997). Questionnaire-based measures are valid for the identification of rural households with hunger and food insecurity. *Journal of Nutrition, 127*, 699–705.

Government Performance and Results Act. (1993). (http://www.whitehouse.gov/omb/mgmt-gpra/gplaw2m.html).

Haberman, S.J. (1977). Maximum likelihood estimation in exponential response models. *Annals of Statistics, 5*, 815–841.

Habicht, J.-P., Pelto, G., Frongillo, E.A., and Rose, D. (2004). *Conceptualization and instrumentation of food insecurity.* Paper presented at the Workshop on the Measurement of Food Insecurity and Hunger, July 15, 2004. Panel to Review the U.S. Department of Agriculture's Measurement of Food Insecurity and Hunger. (http://www7.national academies.org/cnstat/Habicht_etal_paper.pdf).

Haider, S.J. (2005). *A comparison of surveys for food insecurity and hunger measurement*. Background paper prepared for the Panel to Review the U.S. Department of Agriculture's Measurement of Food Insecurity and Hunger. (http://www. national academies/CNSTAT/).

Hamelin, A.M., Beaudry, M., and Habicht, J.-P. (2002). Characterization of household food insecurity in Quebec: Food and feelings. *Social Science and Medicine, 54*, 119–132.

Hamelin, A.M., Habicht, J.-P., and Beaudry, M. (1999). Food insecurity: Consequences for the household and broader social implications. *Journal of Nutrition, 129*, 525–528.

Hamilton, W.L., Cook, J.T., Thompson, W.W., Buron, L.F., Frongillo, E.A., Olson, C.M., and Wehler, C.A. (1997a). *Household food security in the United States in 1995: Summary report of the Food Security Measurement Project*. Alexandria, VA: U.S. Department of Agriculture, Food and Consumer Service. (http://www.fns.usda.gov/oane/MENU/Published/FoodSecurity/SUMRPT.PDF).

Hamilton, W.L., Cook, J.T., Thompson, W.W., Buron, L.F., Frongillo, E.A., Olson, C.M., and Wehler, C.A. (1997b). *Household food security in the United States in 1995: Technical report of the Food Security Measurement Project*. Alexandria, VA: U.S. Department of Agriculture, Food and Consumer Service. (http://www.fns.usda.gov/oane/MENU/Published/FoodSecurity/TECH_RPT.PDF).

Hess, J., and Singer, E. (1995). The role of respondent debriefing questions in questionnaire development. In *1995 Proceedings of the section on survey research methods* (pp. 1075–1080). Washington, DC: American Statistical Association.

Hess, J., Singer, E., and Ciochetto, S. (1996). *Evaluation of the April 1995 Food Security Supplement to the Current Population Survey*. Washington, DC: Center for Survey Methods Research, U.S. Bureau of the Census.

Holben, D.H. (2005). *The concept of hunger*. Background paper prepared for the Panel to Review the U.S. Department of Agriculture's Measurement of Food Insecurity and Hunger. (http://www.nationalacademies/CNSTAT/).

Holland, P.W. (1990). On the sampling theory foundations of item response theory models. *Psychometrika, 55*, 588–601.

Holland, P.W., and Rosenbaum, P.R. (1986). Conditional association and unidimensionality in monotone latent variable models. *Annals of Statistics, 14*, 1523–1543.

Holland, P.W., and Wainer, H. (Eds.). (1993). *Differential item functioning*. Hillsdale, NJ: Erlbaum.

Institute of Medicine. (1986). *Nutrient adequacy: Assessment using food consumption surveys*. Report of the Subcommittee on Criteria for Dietary Evaluation, Coordinating Committee on Evaluation of Food Consumption Surveys, Food and Nutrition Board. Washington, DC: National Academy Press.

Institute of Medicine. (1994). *How should the recommended dietary allowances be revised?* Food and Nutrition Board. Washington, DC: National Academy Press.

Institute of Medicine. (2000). *Dietary reference intakes: Applications in dietary assessment*. Report of the Subcommittee on Interpretation and Uses of Dietary Reference Intakes and the Standing Committee on the Scientific Evaluation of Dietary Reference Intakes, Food and Nutrition Board. Washington, DC: National Academy Press.

Johnson, M.S. (2004). *Item response models and their use in measuring food insecurity and hunger*. Paper presented at the Workshop on the Measurement of Food Insecurity and Hunger, July 15, 2004. Panel to Review the U.S. Department of Agriculture's Measurement of Food Insecurity and Hunger. (http://www7.nationalacademies.org/cnstat/Johnson%20paper.pdf).

Johnson, M.S. (2005). *Methodological issues in measuring food insecurity and hunger*. Background paper prepared for the Panel to Review the U.S. Department of Agriculture's Measurement of Food Insecurity and Hunger. (http://www.nationalacademies.org/CNSTAT/).

Jyoti, D.F., Frongillo, E.A., and Jones, S.J. (2005). Food insecurity affects school children's academic performance, weight gain, and social skills. *Journal of Nutrition, 135*, 2831–2839.

Kelley, T.L. (1923). *Statistical method*. New York: Macmillan.

Kelley, T.L. (1935). *Essential traits of mental life*. Cambridge, MA: Harvard University Press.

Kleinman, R.E., Murphy, J.M., Little, M., Pagano, M., Wehler, C.A., Regal, K., and Jellinek M.S. (1998). Hunger in children in the United States: Potential behavioral and emotional correlates. *Pediatrics, 101*, 3.

Laitinen, J., Power, C., and Jarvelin, M.R. (2001). Family social class, maternal body mass index, childhood body mass index, and age at menarche as predictors of adult obesity. *American Journal of Clinical Nutrition, 74*(3), 287–294.

Lawley, D.N. (1943). On problems connected with item selection and test construction. *Proceedings of the Royal Society of Edinburgh, 61*, 273–287.

Lazarsfeld, P.F., and Henry, N.W. (1968). *Latent structure analysis*. New York: Houghton-Mifflin.

LeBlanc, M., Kuhn, B., and Blaylock, J. (2005). Poverty amidst plenty: Food insecurity in the United States. In *Reshaping agriculture's contributions to society: Proceedings of the 25th international conference of agricultural economists*. London: Blackwell Publishing.

Lee, J.S., and Frongillo, EA. (2001a). Nutritional and health consequences are associated with food insecurity among U.S. elderly persons. *Journal of Nutrition, 131*, 1503–1509.

Lee, J.S., and Frongillo, E.A. (2001b). Factors associated with food insecurity among U.S. elderly: Importance of functional impairments. *Journal of Gerontology: Social Sciences, 56B*, S94–S99.

Lindsey, B., Clogg, C.C., and Grego, J. (1991). Semiparametric estimation in the Rasch model and related exponential response models, including a simple latent class model for item analysis. *Journal of the American Statistical Association, 86*, 96–107.

Lord, F.M. (1952). *A theory of test scores*. Psychometric monograph no. 7. Iowa City, IA: The Psychometric Society.

Lord, F.M. (1980). *Application of item response theory to practical testing problems*. Hillsdale, NJ: Erlbaum.

Lord, F.M., and Novick, M.R. (1968). *Statistical theories of mental test scores*. Reading, MA: Addison-Wesley.

Masters, G.N. (1982). A Rasch model for partial credit scoring. *Psychometrika, 47*, 149–174.

Maxwell, S., and Frankenberger, T. (1992). *Household food security: Concepts, indicators, measurements. A technical review*. New York and Rome: United Nations Children's Fund and International Fund for Agricultural Development.

Mislevy, R.J., Beaton, A.E., Kaplan, B., Sheehan, K.M., et al. (1992). Estimating population characteristics from sparse matrix samples of item responses. *Journal of Educational Measurement, 29*, 133–162.

Muraki, E. (1992). A generalized partial credit model: Application of an EM algorithm. *Applied Psychological Measurement, 16*, 159–176.

Murphy, J.M., Wehler, C.A., Pagano, M.E., et al. (1998). Relationships between hunger and psychosocial functioning in low-income American children. *Journal of the American Academy of Child and Adolescent Psychiatry, 37*, 163–171.

Naiken, L. (2003). FAO methodology for estimating the prevalence of undernourishment. In *Proceedings of international scientific symposium on measurement and assessment of food deprivation and undernutrition* (pp. 7–47). June 26–28, 2002. Rome: Food and Agriculture Organization of the United Nations.

National Research Council. (2005). *Measuring literacy: Performance levels for adult literacy*. Board on Testing and Assessment, Center for Education, Division of Behavioral and Social Sciences and Education. Washington, DC: The National Academies Press.

Nord, M. (1999, January 12). *Upward bias on food insecurity and hunger prevalence estimates due to measurement error*. Unpublished staff working paper #FS-8. U.S. Department of Agriculture, Economic Research Service.

Nord, M. (2002a). Food security rates are high for elderly households. *Food Review, 25*(2), 19–24.

Nord, M. (2002b). *A 30-day food security scale for Current Population Survey Food Security Supplement data*. E-FAN-02-015. Washington, DC: U.S. Department of Agriculture, Economic Research Service. (http://www.ers.usda/publications/efan02015/).

Nord, M., Andrews, M., and Carlson, S. (2004). *Household food security in the United States, 2003*. Food Assistance and Nutrition Research Report No. 42. Washington, DC: U.S. Department of Agriculture, Economic Research Service.

Nord, M., Andrews, M., and Carlson, S. (2005a). *ERS and FNS responses to measuring food insecurity and hunger: Phase 1 report*. Presentation at meeting on March 28, 2005, of the Panel to Review the U.S. Department of Agriculture's Measurement of Food Insecurity and Hunger. Washington, DC.

Nord, M., Andrews, M., and Carlson, S. (2005b). *Household food security in the United States, 2004*. Economic Research Report No. 11. Washington, DC: U.S. Department of Agriculture, Economic Research Service.

Nord, M., Andrews, M., and Winicki, J. (2002a). Frequency and duration of food insecurity and hunger in U.S. households. *Journal of Nutrition Education and Behavior, 34*, 194–201.

Nord, M., and Bickel, G. (2002). *Measuring children's food security in U.S. households, 1995–1999*. Food Assistance and Nutrition Research Report No. 25. Washington, DC: U.S. Department of Agriculture, Economic Research Service. (http://www.ers.usda.gov/publications/fanrr25).

Nord, M., Sathpathy, A.K., Raj, N., Webb, P., and Houser, R. (2002b). *Comparing household survey-based measures of food insecurity across countries: Case studies in India, Uganda, and Bangladesh*. Discussion Paper No. 7. Friedman School of Nutrition Science, Tufts University. (http://nutrition.Tufts.edu/publications.fpan).

Ohls, J., Prakash, A., Radbill, L., and Schirm, A. (1999). *Methodological findings and early conclusions based on the 1995, 1996 and 1997 food security data*. Unpublished paper, January 5, 1999. Mathematica Policy Research, Inc.

Ohls, J., Radbill, L., and Schirm, A. (2001). *Household food security in the United States, 1995–1997: Technical issues and statistical report*. Final Report of the Project to Analyze 1996 and 1997 Food Security Data. Alexandria, VA: U.S. Department of Agriculture, Food and Nutrition Service.

Olson, C.M. (1999). Symposium: Advances in measuring food insecurity and hunger in the U.S. Introduction. *Journal of Nutrition, 129*(2), 504S–505S.

Opsomer, J.D., Jensen, H.H., and Pan, S. (2003). An evaluation of the USDA food security measure with generalized linear mixed models. *Journal of Nutrition, 133*, 421–427. Errata, *Journal of Nutrition, 133*, 2394.

Opsomer, J.D., Jensen, H.H., Nusser, S.M., Drignel, D., and Amemiya, Y. (2002). *Statistical considerations for the USDA Food Security Index*. Working Paper 02-WP 307. Center for Agricultural and Rural Development, Iowa State University.

Parzen, E. (1960). *Modern probability theory and its applications*. New York: Wiley.

Patz, R., and Junker, B.W. (1999). Applications and extensions of MCMC in IRT: Multiple item types, missing data, and rated responses. *Journal of Educational and Behavioral Statistics, 24*, 342–366.

Pelletier, D.L., Olson, C.M., and Frongillo, E.A. (2001). Food insecurity, hunger and undernutrition. Chapter 60. In B. Bowman and R. Russell (Eds.), *Present knowledge in nutrition, eighth edition* (pp. 698–710). Washington, DC: International Life Sciences Institute Press.

Pelto, P.J., and Pelto, G.H. (1978). *Anthropological research: The structure of inquiry*. New York: Cambridge University Press.

Pérez-Escamilla, R., Segall-Corrêa, A.M., Maranha, L.K., Sampaio, M.A., Marin-León L., and Panigassi, G. (2004). An adapted version of the U.S. Department of Agriculture food insecurity module is a valid tool for assessing household food insecurity in Campinas, Brazil. *Journal of Nutrition, 134*, 1923–1928.

Peterson, W.W., Birdsall, T.G., and Fox, W.C. (1954). The theory of signal detectability. *IRE Transactions of the Professional Group on Information Theory, 4*, 171–212.

Presser, S., Rothgeb, J.M., Couper, M.P., et al. (Eds.). (2004). *Methods for testing and evaluating survey questionnaires*. New York: John Wiley and Sons.

Quandt, S.A., Arcury, T.A., McDonald, J., Bell, R.A., and Vitolins, M.Z. (2001). Meaning and management of food security among rural elders. *Journal of Applied Gerontology, 20*(3), 356–376.

Quandt, S.A., McDonald, J., Arcury, T.A., Bell, R.A., and Vitolins, M.Z. (2000). Nutritional self-management of elderly widows in rural communities. *Gerontologist, 40*(1), 86–96.

Quandt, S.A., and Rao, P. (1999). Hunger and food security among older adults in a rural community. *Human Organization, 58*(1), 28–35.

Radimer, K.L., Olson, C.M., and Campbell, C.C. (1990). Development of indicators to assess hunger. *Journal of Nutrition, 120*, 1544–1548.

Radimer, K.L., Olson, C.M., Greene, J.C., Campbell, C.C., and Habicht, J.-P. (1992). Understanding hunger and developing indicators to assess it in women and children. *Journal of Nutrition Education, 24*, 36S–45S.

Ramsay, J.O. (1973). The effect of number of categories in rating scales on precision of estimation of scale values. *Psychometrika, 28*, 513–532.

Rasch, G. (1960). *Probabilistic models for some intelligence and attainment tests*. Copenhagen: Danmarks Paedagogiske Institut.

Reid, L. (2001, January). *The consequences of food insecurity for child well-being: An analysis of children's school achievement, psychological well-being, and health*. Joint Center for Poverty Research, Working Paper 137.

Rose, D. (1997). *Assessing food insecurity in the United States: Background information for domestic follow-up activities to the World Food Summit*. ERS Staff Paper No. 9706. Washington, DC: U.S. Department of Agriculture, Economic Research Service, Food and Consumer Economics Division.

Rose, D., Basiotis, P.P., and Klein, B.W. (1995). Improving federal efforts to assess hunger and food insecurity. *Food Review, 18*, 18–23.

Rose, D., Habicht, J.-P., and Devaney, B. (1998). Household participation in the food stamp and WIC programs increases the nutrient intakes of preschool children. *Journal of Nutrition, 128*, 548–555.

Rose, D., and Oliveira, V. (1997). Nutrient intakes of individuals from food-insufficient households in the United States. *American Journal of Public Health, 87*, 1956–1961.

Ross, M., and Conway, M. (1986). Remembering one's own past: The construction of personal histories. In R. Sorrentino and E.T. Higgins (Eds.), *Handbook of motivation and cognition: Foundations of social behavior*. Chichester, U.K.: John Wiley and Sons.

Ruel, M.T., Rivera, J., Habicht, J.-P., and Martorell, R. (1995). Differential response to early nutrition supplementation: Long-term effects on height at adolescence. *International Journal of Epidemiology, 24*, 404–412.

Samejima, F. (1969). *Estimation of latent ability using a response pattern of graded scores*. Psychometric Monograph No. 17. Iowa City, IA: The Psychometric Society.

Schaeffer, N., and Presser, S. (2003), The science of asking questions. *Annual Review of Sociology, 29*, 65–88.

Schwarz, N. (1994). Judgment in social context: Biases, shortcomings, and the logic of conversation. *Advances in Experimental Social Psychology, 26*, 123–162.

Shook Slack, K., and Yoo, J. (2004). *Food hardships and child behavior problems among low-income children.* Discussion Paper No. 1290–04. Madison, WI: Institute for Research on Poverty.

Siefert, K., Heflin, C.M., Corcoran, M.E., and Williams, D.R. (2004, June). Food insufficiency and physical and mental health in a longitudinal survey of welfare recipients. *Journal of Human and Social Behavior, 45*, 171–186.

Smith, L. (1998). Can FAO's measure of chronic undernourishment be strengthened? *Food Policy, 23*(5), 425–445.

Spearman, C. (1904). "General intelligence" objectively determined and measured. *American Journal of Psychology, 15*, 210–292.

Spearman, C. (1907). Demonstration of formulae for true measurement correlation. *American Journal of Psychology, 18*, 161–169.

Stormer, A., and Harrison, G.G. (2003). *Does household food security affect cognitive and social development of kindergarteners?* Discussion Paper No. 1276–03. Madison, WI: Institute for Research on Poverty.

Stouffer, S.A., Guttman, L., Suchman, E.A., Lazarsfeld, P.F., et al. (1950). *Measurement and prediction.* Princeton, NJ: Princeton University Press.

Stout, W., Habing, B., Douglas, J., et al. (1996). Conditional covariance-based nonparametric multidimensionality assessment. *Applied Psychological Measurement, 20*, 331–354.

Thurstone, L.L. (1931). Multiple factor analysis. *Psychological Review, 38*, 406–427.

Tourangeau, R., Rips, L., and Rasinski, K. (2000). *The psychology of survey response.* Cambridge, UK: Cambridge University Press.

Townsend, M., Peerson, J., Love, B., Achterberg, C., and Murphy, S. (2001). Food insecurity is positively related to overweight in women. *Journal of Nutrition, 131*, 1738–1745.

Tucker, L.R. (1946). Maximum validity of a test with equivalent items. *Psychometrika, 11*, 1–14.

U.S. Department of Agriculture. (1995). *Food security measurement and research conference: Papers and proceedings.* Alexandria, VA: U.S. Department of Agriculture, Food and Consumer Service, Office of Analysis and Evaluation.

U.S. Department of Agriculture. (2004). *Characteristics of food stamp households: Fiscal year 2003.* FSP-04-CHAR, by K. Cunnyngham and B. Brown, Mathematica Policy Research, Inc. Alexandria, VA: U.S. Department of Agriculture, Food and Nutrition Service, Office of Analysis.

U.S. Department of Agriculture and U.S. Department of Health and Human Services. (1994). *Conference on food security measurement and research: Papers and proceedings.* January 21–22, 1994. Washington, DC.

U.S. General Accounting Office. (1998). *Performance measurement and evaluation: Definitions and relationships.* GAO/GGD-98-26. Washington, DC.

U.S. Government Accountability Office. (2002). Reports on the Government Performance and Results Act of 1993 (PL 103–62). (http://www.gao.gov/new.items/gpra/gpra.htm)

U.S. President, Task Force on Food Assistance. (1984). *Report of the President's Task Force on Food Assistance.* Washington, DC: U.S. Government Printing Office.

Wang, X., Bradlow, E.T., and Wainer, H. (2002). A general Bayesian model for testlets: Theory and applications. *Applied Psychological Measurement, 26*(1), 109–128.

Webb, P., Coates, J., and Houser, R. (2003). Challenges in defining direct measures of hunger and food insecurity in Bangladesh: Findings from ongoing fieldwork. In *Proceedings of international scientific symposium on measurement and assessment of food deprivation and undernutrition* (pp. 301–303). June 26–28, 2002. Rome: Food and Agriculture Organization of the United Nations.

Wehler, C.A., Scott, R.I., and Anderson, J.J. (1992). The Community Childhood Hunger Identification Project: A model of domestic hunger—Demonstration project in Seattle, Washington. *Journal of Nutrition Education, 24,* 29S–35S.

Weinreb, L., Wehler, C., Perloff, J., et al. (2005). Hunger: Its impact on children's health and mental health. *Pediatrics, 110,* 41–50.

Wilde, P. (2004). *The uses and purposes of the USDA food security and hunger measure.* Paper presented at the Workshop on the Measurement of Food Insecurity and Hunger, July 15, 2004. Panel to Review the U.S. Department of Agriculture's Measurement of Food Insecurity and Hunger. (http://www7.nationalacademies.org/cnstat/Wilde_paper_tables.pdf).

Wilde, P., and Nord, M. (2005). The effect of food stamps on food security: A panel approach. *Review of Agricultural Economics, 27*(3), 425–432.

Willis, G. (2004). *Cognitive interviewing: A tool for improving questionnaire design.* Thousand Oaks, CA: Sage Publications.

Winicki, J., and Jemison, K. (2003). Food insecurity and hunger in the kindergarten classroom: Its effect on learning and growth. *Contemporary Economic Policy, 21,* 145–157.

Wolfe, W.S., Frongillo, E.A., and Valois, P. (2003). Understanding the experience of elderly food insecurity suggests ways to improve its measurement. *Journal of Nutrition, 133,* 2762–2769.

Wolfe, W.S., Olson, C.M., Kendal, A., and Frongillo, E.A. (1996). Understanding food insecurity in the elderly: A conceptual framework. *Journal of Nutrition Education, 28*(2), 92–100.

Wolfe, W.S., Olson, C.M., Kendal, A., and Frongillo, E.A. (1998). Hunger and food insecurity in the elderly: Its nature and measurement. *Journal of Aging and Health, 10*(3), 327–350.

Acronyms and Abbreviations

CCHIP	Community Childhood Hunger Identification Project
CNSTAT	Committee on National Statistics
CPS	Current Population Survey
CSMR	Center for Survey Methods Research (U.S. Census Bureau)
DHHS	U.S. Department of Health and Human Services
DIF	Differential item functioning
ECLS-B	Early Childhood Longitudinal Survey, Birth Cohort of 2002
ECLS-K	Early Childhood Longitudinal Survey, Kindergarten Class of 1998–1999
ERS	Economic Research Service (U.S. Department of Agriculture)
FAO	United Nations Food and Agricultural Organization
FNS	Food and Nutrition Service (U.S. Department of Agriculture)
FSP	Food Stamp Program
FSS	Food Security Supplement
GPRA	Government Performance and Results Act
HFSSM	Household Food Security Survey Module
IRT	Item response theory

LSRO   Life Sciences Research Office (Federation of American Societies for Experimental Biology)

MPR   Mathematica Policy Research, Inc.

NAAL   National Assessment of Adult Literacy
NAEP   National Assessment of Educational Progress
NCHS   National Center for Health Statistics
NHANES   National Health and Nutrition Examination Survey
NHIS   National Health Interview Survey
NNMRRP   National Nutrition Monitoring and Related Research Program

RDA   Recommended Dietary Allowances

SIPP   Survey of Income and Program Participation
SPD   Survey of Program Dynamics

USDA   U.S. Department of Agriculture

WIC   Special Supplemental Nutrition Program for Women, Infants, and Children

Appendix A

Current Population Survey Food Security Supplement Questionnaire, December 2003

[Not all follow-up questions have been numbered separately]

I. FOOD EXPENDITURES

These first questions are about all the places at which you bought food LAST WEEK. By LAST WEEK, I mean from Sunday through Saturday.

1. First, did (you/anyone in your household) shop for food at a supermarket or grocery store LAST WEEK?

2. Think about other places where people buy food, such as meat markets, produce stands, bakeries, warehouse clubs, and convenience stores. Did (you/anyone in your household) buy food from any stores such as these LAST WEEK?

3. LAST WEEK, did (you/anyone in your household) buy food at a restaurant, fast food place, cafeteria, or vending machine? (Include any children who may have bought food at the school cafeteria).

4. Did (you/anyone in your household) buy food from any other kind of place LAST WEEK?

Now I'm going to ask you about the ACTUAL amount you spent on food LAST WEEK in all the places where you bought food. Then, since LAST WEEK may have been unusual for you, I will ask about the amount you USUALLY spend.

5. How much did (you/your household) ACTUALLY spend at supermarkets and grocery stores LAST WEEK (including any purchases made with food stamps)?

> How much of the (amount from last question) was for nonfood items, such as pet food, paper products, detergents, or cleaning supplies?

6. How much did (you/your household) spend at stores such as meat markets, produce stands, bakeries, warehouse clubs, and convenience stores LAST WEEK (including any purchases made with food stamps)?

> How much of the (amount from last question) was for nonfood items, such as pet food, paper products, detergents, or cleaning supplies?

7. How much did (you/your household) spend for food at restaurants, fast food places, cafeterias, and vending machines LAST WEEK?

8. How much did (you/your household) spend for food at any other kind of place LAST WEEK?

(Let's see, it seems that (you/your household) did not buy any food LAST WEEK. Let's see, (you/your household) spent about (fill with $80) on food LAST WEEK.) Now think about how much (you/your household) USUALLY (spend/spends). How much (do you/does your household) USUALLY spend on food at all the different places we've been talking about IN A WEEK? (Please include any purchases made with food stamps). Do not include nonfood items such as pet food, paper products, detergent or cleaning supplies.

II. MINIMUM SPENDING NEED TO HAVE ENOUGH FOOD

9. In order to buy just enough food to meet (your needs/the needs of your household), would you need to spend more than you do now, or could you spend less?

10. About how much MORE would you need to spend each week to buy just enough food to meet the needs of your household?

11. About how much LESS could you spend each week and still buy enough food to meet the needs of your household?

III. FOOD PROGRAM PARTICIPATION

People do different things when they are running out of money for food in order to make their food or their food money go further.

12. In the last 12 months, since December of last year, did you ever run short of money and try to make your food or your food money go further?

13. In the past 12 months, since December of last year, did (you/anyone in this household) get food stamp benefits, that is, either food stamps or a food-stamp benefit card?

14. In which months of 2003 were food stamps received?

15. On what date in November did (you/your household) receive food stamp benefits?

16. How much did (you/your household) receive the last time you got food stamp benefits?

17. During the past 30 days, did (your child/any children in the household between 5 and 18 years old) receive free or reduced-cost lunches at school?

18. During the past 30 days, did (your child/any children in the household) receive free or reduced-cost breakfasts at school?

19. During the past 30 days, did (your child/any children in the household) receive free or reduced-cost food at a day-care or Head Start program?

20. During the past 30 days, did any (women/women or children/children/ women and children) in this household get food through the WIC program?

21. How many (women/women or children/children/women and children) in the household got WIC foods?

IV. FOOD SUFFICIENCY AND FOOD SECURITY

The next questions are about the food eaten in your household in the last 12 months, since December of last year, and whether you were able to afford the food you need.

22. Which of these statements best describes the food eaten in your household—enough of the kinds of food we want to eat, enough but not always the *kinds* of food we want to eat, sometimes not enough to eat, or often not enough to eat?

Now I'm going to read you several statements that people have made about their food situation. For these statements, please tell me whether the statement was OFTEN true, SOMETIMES true, or NEVER true for (you/ your household) in the last 12 months.

23. The first statement is "(I/We) worried whether (my/our) food would run out before (I/we) got money to buy more." Was that OFTEN true, SOMETIMES true, or NEVER true for (you/your household) in the last 12 months?
    Did this ever happen in the last 30 days?

24. "The food that (I/we) bought just didn't last, and (I/we) didn't have money to get more." Was that OFTEN, SOMETIMES or NEVER true for you in the last 12 months?
    Did this ever happen in the last 30 days?

25. "(I/we) couldn't afford to eat balanced meals." Was that often, sometimes or never true for you in the last 12 months?
    Did this ever happen in the last 30 days?

26. "(I/we) relied on only a few kinds of low-cost food to feed ((my/our) child/the children) because (I was/we were) running out of money to buy food. Was that often, sometimes or never true for you in the last 12 months?
    Did this ever happen in the last 30 days?

27. "(I/we) couldn't feed ((my/our) child/the children) a balanced meal, because (I/we) couldn't afford that." Was that often, sometimes, or never true for you in the last 12 months?
    Did this ever happen in the last 30 days?

28. "((My/Our) child was/The children were) not eating enough because (I/ we) just couldn't afford enough food." Was that often, sometimes or never true for you in the last 12 months?
    Did this ever happen in the last 30 days?

29. In the last 12 months, did you or other adults in your household ever cut the size of your meals or skip meals because there wasn't enough money for food?
    How often did this happen—almost every month, some months but not every month, or in only 1 or 2 months?

30. Now think about the last 30 days. During that time did (you/you or other adults in your household) ever cut the size of your meals or skip meals because there wasn't enough money for food?

How many days did this happen in the last 30 days?

31. In the last 12 months, did you ever eat less than you felt you should because there wasn't enough money for food?

How often did this happen—almost every month, some months but not every month, or in only 1 or 2 months?

Did this happen in the last 30 days?

32. In the last 30 days, how many days did you eat less than you felt you should because there wasn't enough money to buy food?

33. In the last 12 months, since December of last year, were you ever hungry but didn't eat because you couldn't afford enough food?

How often did this happen—almost every month, some months but not every month, or in only 1 or 2 months?

Did this happen in the last 30 days?

34. In the last 30 days, how many days were you hungry but didn't eat because you couldn't afford enough food?

35. In the last 12 months, did you lose weight because you didn't have enough money for food?

Did this happen in the last 30 days?

36. In the last 12 months, since last December, did (you/you or other adults in your household) ever not eat for a whole day because there wasn't enough money for food?

How often did this happen—almost every month, some months but not every month, or in only 1 or 2 months?

37. Now think about the last 30 days. During that time did (you/you or other adults in your household) ever not eat for a whole day because there wasn't enough money for food?

How many times did this happen in the last 30 days?

The next questions are about (your child/children living in the household who are under 18 years old).

38. In the last 12 months, since December of last year, did you ever cut the size of (your child's/any of the children's) meals because there wasn't enough money for food?
     How often did this happen—almost every month, some months but not every month, or in only 1 or 2 months?
     Did this happen in the last 30 days?

39. In the last 30 days, how many days did you cut the size of (your child's/ the children's) meals because there wasn't enough money for food?

40. In the last 12 months, (was your child/were the children) ever hungry but you just couldn't afford more food?
     How often did this happen—almost every month, some months but not every month, or in only 1 or 2 months?
     Did this happen in the last 30 days?

41. In the last 30 days, how many days (was your child/were the children) hungry but you just couldn't afford more food?

42. In the last 12 months, did (your child/ any of the children) ever skip a meal because there wasn't enough money for food?
     How often did this happen—almost every month, some months but not every month, or in only 1 or 2 months?

43. Now think about the last 30 days. Did (your child/the children) ever skip a meal during that time because there wasn't enough money for food?
     How many days did this happen in the last 30 days?

44. In the last 12 months, since December of last year, did (your child/any of the children) ever not eat for a whole day because there wasn't enough money for food?
     Did this happen in the last 30 days?

V. WAYS OF COPING WITH NOT HAVING ENOUGH FOOD

45. During the past 30 days, did (you/anyone in the household) receive any meals delivered to the home from community programs, "Meals on Wheels," or any other programs?

46. During the past 30 days, did (you/anyone in the household) go to a community program or senior center to eat prepared meals?

47. In the last 12 months, did (you/you or other adults in your household) ever get emergency food from a church, a food pantry, or food bank?
    How often did this happen—almost every month, some months but not every month, or in only 1 or 2 months?
    Did this happen in the last 30 days?

48. Is there a church, food pantry or food bank in your community where you could get emergency food if you needed it?

49. In the last 12 months, did (you/you or other adults in your household) ever eat any meals at a soup kitchen?
    How often did this happen—almost every month, some months but not every month, or in only 1 or 2 months?
    Did this happen in the last 30 days?

Appendix B

Biographical Sketches of
Panel Members and Staff

JANET L. NORWOOD *(Chair)* is a counselor and senior fellow at The Conference Board, where she chairs the Advisory Committee on the Leading Indicators. She served as U.S. Commissioner of Labor Statistics from 1979 to 1992 and then was a senior fellow at the Urban Institute until 1999. She is a past member of the Committee on National Statistics and the Division of Engineering and Physical Sciences of the National Research Council. She currently serves on the Board of Scientific Counselors at the National Center for Health Statistics and was designated a National Associate of the National Research Council in 2001. She is a fellow and past president of the American Statistical Association, a member and past vice president of the International Statistical Institute, an honorary fellow of the Royal Statistical Society, and a fellow of the National Academy of Public Administration and the National Association of Business Economists. She has a B.A. from Rutgers University and M.A. and Ph.D. degrees from the Fletcher School of Law and Diplomacy of Tufts University. She has received honorary LL.D. degrees from Carnegie Mellon, Florida International, Harvard, and Rutgers universities.

ERIC T. BRADLOW is the K.P. Chao professor of marketing and statistics at the Wharton School of the University of Pennsylvania. He also serves as associate editor for the *Journal of Computational and Graphical Statistics* and *Psychometrika* and as senior associate editor for the *Journal of Educational and Behavioral Statistics*. He has won numerous teaching awards and his research interests include Bayesian modeling, statistical computing, and developing new methodology for unique data structures. His current projects center on optimal resource allocation, choice modeling, and complex latent structures. He has a Ph.D. in mathematical statistics from Harvard University.

J. MICHAEL BRICK is senior statistician, vice president, and associate director of the statistical staff at Westat. He has 25 years of experience and expertise in sample design and estimation for large surveys, the theory and practice of telephone surveys, the techniques of total quality management and survey quality control, nonresponse and bias evaluation, and survey methodology. He has contributed to the statistical and substantive aspects of numerous studies and to statistical methodology research in several areas, including education, transportation, and product injury studies. He is a fellow of the American Statistical Association, an elected member of the International Statistical Institute, and a research professor in the Joint Program in Survey Methodology at the University of Maryland. He has a B.S. in mathematics from the University of Dayton and M.A. and Ph.D. degrees in statistics from American University.

EDWARD A. FRONGILLO, JR., is associate professor in the Division of Nutritional Sciences and director of the Office of Statistical Consulting at Cornell University. He previously served as the director of the Program in International Nutrition at Cornell. His current research activities include the Multicentre Growth Reference Study of the World Health Organization; the conceptualization, measurement, and consequences of food insecurity in elders and families in poor countries and in North America; and the role of food assistance programs in alleviating consequences of food insecurity. He is a member of the editorial boards of the *Journal of Nutrition* and the *Journal of Gerontology: Medical Sciences.* He has an M.S. in biometry, an M.S. in human nutrition, and a Ph.D. in biometry, all from Cornell University.

PAUL W. HOLLAND holds the Frederic M. Lord chair in measurement and statistics at the Educational Testing Service (ETS). His association with ETS began in 1975 as director of the Research Statistics Group, and in 1986 he was appointed its first distinguished research scientist. He left ETS in 1993 to join the faculty at University of California, Berkeley, as a professor in the Graduate School of Education and Department of Statistics but returned in 2000 to his current position at ETS. His research interests include psychometrics, causal inference of educational interventions in nonexperimental studies, discrete multivariate data analysis, and the analysis of social networks. He was designated a national associate of the National Research Council in 2002. He has an M.A. and a Ph.D. in statistics from Stanford University (1966) and a B.A. in mathematics from the University of Michigan (1962).

MICHAEL D. HURD is a senior economist and the director for the RAND Center for the Study of Aging. His expertise concerns aging and the elderly; savings, wealth, and retirement; and U.S. labor markets and social security. Previously he chaired the Department of Economics at the State University of New York at Stony Brook. He was a visiting senior scientist at the Institute for Social Research at the University of Michigan and a visiting associate professor of economics at Stanford University. He is a member of the Behavior and Sociology of Aging Review Subcommittee at the National Institutes of Health. He is also a member of the Scientific Committee of the Center for Research on Pensions and Welfare Policies at the University of Turin, Italy. He is a consultant to the English Longitudinal Study of Aging and a consultant to the Survey on Health, Aging, and Retirement in Europe. He has a Ph.D. in economics from the University of California, Berkeley.

HELEN H. JENSEN is professor of economics and head of the Center for Agricultural and Rural Development's food and nutrition policy research division at Iowa State University. Her research addresses food assistance and nutrition policies, food security and the economics of food safety and food hazard control options. She is on the editorial boards of *Agricultural Economics, Food Economics,* and *Agribusiness: An International Journal* and was elected chair of the Food Safety and Nutrition Section of the American Agricultural Economics Association. She is currently serving on the Institute of Medicine's Committee to Review the WIC Food Packages and the National Research Council's Committee on Assessing the Nation's Framework for Addressing Animal Diseases. She has been a member of the National Research Council's panel on animal health and food safety and expert panels related to food safety, food insecurity and hunger, and food programs. She has a Ph.D. in agricultural economics from the University of Wisconsin-Madison.

NANCY MATHIOWETZ is associate professor of sociology at the University of Wisconsin-Milwaukee. She was previously an associate professor at the University of Maryland's Joint Program in Survey Methodology. Her research interests include the assessment and reduction of measurement error in surveys and the use of survey data in the development of public policy. She is co-editor of *Survey Measurement of Work Disability: Summary of a Workshop*, one of the reports of the Committee to Review the Social Security Administration's Disability Decision Process Research, a joint project of the Institute of Medicine and the National Research Council. She serves as associate editor of *Public Opinion Quarterly* and the *Jour-*

nal of Official Statistics. She has an M.S. in biostatistics and a Ph.D. in sociology, both from the University of Michigan.

SUSAN E. MAYER is dean and associate professor at the Harris Graduate School of Public Policy Studies and the College at the University of Chicago. She also serves as a faculty affiliate with the University's Center for Human Potential and Public Policy. She is past director of the Northwestern University/University of Chicago Joint Center for Poverty Research. Her current research is on the effect of economic mobility across generations and the role of noncognitive skills on social and economic success. She is author of the book, *What Money Can't Buy: Family Income and Children's Life Chances*, and co-editor of *Earning and Learning: How Schools Matter*. She has a Ph.D. in sociology from Northwestern University.

DONALD DIEGO ROSE is associate professor in the Department of Community Health Science at Tulane University. His research focuses on the determinants of food consumption in low-income populations, and on food programs and nutrition policies in both domestic and international contexts. Previously he was project director/nutritionist for the WIC nutrition program in a farmworker clinic in rural California, as well as a research team leader with the U.S. Department of Agriculture's Economic Research Service, studying the determinants and consequences of household food insecurity in America, the nutrition and health impacts of food assistance programs, and the evaluation of low-income nutrition education projects. He also worked on food security and nutrition issues in Mozambique with Michigan State University's Food Security Project and in South Africa with the University of Cape Town's Medical School. He has an M.P.H. in public health nutrition and a Ph.D. in agricultural economics from the University of California, Berkeley.

GOOLOO S. WUNDERLICH (*Study Director*) is a member of the staff of the Committee on National Statistics. She has 48 years of experience at the program and policy levels in health and population policy analysis, research, and statistics in the U.S. Public Health Service, President's Advisory Commission on Rural America, the U.S. Census Bureau, and the National Academies. Her professional interests and experience have focused on the conduct and analysis of national health surveys, analysis and public policy formulation relating to population research, family planning, aging, long-term care, and a wide range of health policy issues. At the National Academies, she has served as study director for projects on the future of rural health, the Social Security Administration's disability decision process research, improving the quality of long-term care, the adequacy of nurse staffing in hospitals and nursing homes, and the National Health Care Survey.

She is a member of the National Academy of Social Insurance. She has B.A., M.A., and Ph.D. degrees from the University of Bombay, India, and completed two years of postdoctoral studies at the University of Minnesota and the University of Chicago.

Index